# CAMBRIDGE LIBRARY COLLECTION

*Books of enduring scholarly value*

## Darwin

Two hundred years after his birth and 150 years after the publication of 'On the Origin of Species', Charles Darwin and his theories are still the focus of worldwide attention. This series offers not only works by Darwin, but also the writings of his mentors in Cambridge and elsewhere, and a survey of the impassioned scientific, philosophical and theological debates sparked by his 'dangerous idea'.

## Catalogue of the Library of Charles Darwin Now in the Botany School, Cambridge

For those engaged in research on Darwin or his circle, the Darwin Library (now housed at the Cambridge University Library) is a treasure trove. It contains handwritten scribbles on book pages, note-filled scraps of paper pinned to back covers and personal inscriptions from mentors such as J. S. Henslow. It provides fascinating insights into Darwin's own reading, for example that he read Patrick Matthew's On Naval Timber and Arboriculture, dated 1831, which speaks of the 'natural process of selection among plants', long before he published The Origin of Species (1859). Published in 1908, this catalogue to the Darwin Library collection, with an introduction by Darwin's son Francis, is a gateway into Darwin's thought, research and intellectual context, via his personal books and pamphlets. The book lists works in English and other languages, and contains bibliographic information, including the original publisher and date of publication, together with details of translations.

Cambridge University Press has long been a pioneer in the reissuing of out-of-print titles from its own backlist, producing digital reprints of books that are still sought after by scholars and students but could not be reprinted economically using traditional technology. The Cambridge Library Collection extends this activity to a wider range of books which are still of importance to researchers and professionals, either for the source material they contain, or as landmarks in the history of their academic discipline.

Drawing from the world-renowned collections in the Cambridge University Library, and guided by the advice of experts in each subject area, Cambridge University Press is using state-of-the-art scanning machines in its own Printing House to capture the content of each book selected for inclusion. The files are processed to give a consistently clear, crisp image, and the books finished to the high quality standard for which the Press is recognised around the world. The latest print-on-demand technology ensures that the books will remain available indefinitely, and that orders for single or multiple copies can quickly be supplied.

The Cambridge Library Collection will bring back to life books of enduring scholarly value across a wide range of disciplines in the humanities and social sciences and in science and technology.

# Catalogue of the Library of Charles Darwin now in the Botany School, Cambridge

*Compiled by H. W. Rutherford, of the University Library; with an Introduction by Francis Darwin*

EDITED BY HENRY WILLIAM RUTHERFORD

CAMBRIDGE UNIVERSITY PRESS

Cambridge New York Melbourne Madrid Cape Town Singapore São Paolo Delhi

Published in the United States of America by Cambridge University Press, New York

www.cambridge.org
Information on this title: www.cambridge.org/9781108002363

© in this compilation Cambridge University Press 2009

This edition first published 1908
This digitally printed version 2009

ISBN 978-1-108-00236-3

Catalogue of the Library of
Charles Darwin now in the
Botany School, Cambridge

CAMBRIDGE UNIVERSITY PRESS WAREHOUSE

C. F. CLAY, Manager.

London: FETTER LANE, E.C.

Glasgow: 50, WELLINGTON STREET.

ALSO

London: H. K. LEWIS, 136, GOWER STREET, W.C.

Leipzig: F. A. BROCKHAUS.

New York: G. P. PUTNAM'S SONS.

Bombay and Calcutta: MACMILLAN AND CO., Ltd.

# Catalogue of the Library of Charles Darwin now in the Botany School, Cambridge

Compiled by H. W. Rutherford, of the University Library; with an Introduction by Francis Darwin.

CAMBRIDGE :
at the University Press
1908

**Cambridge:**
PRINTED BY JOHN CLAY, M.A
AT THE UNIVERSITY PRESS.

# PREFACE.

IN issuing this catalogue of the Library of Charles Darwin I wish to express to Mr Francis Darwin on behalf of the Staff of the Botany School our keen appreciation of his generosity. Rather more than a year ago Mr Darwin spoke to me of his intention to transfer the whole of his father's Library to the Botany School with the view of bequeathing it to the University. The books have been placed in a set of locked cases apart from the general departmental Library. From a purely historical point of view the Darwin Library is obviously of great interest, while the numerous notes scattered through many of the volumes add considerably to its scientific value. For the convenience of those engaged in research and as a means of adding to our knowledge of Darwin's scientific life, it seemed desirable to publish a catalogue of the books and pamphlets. The work has been done by Mr Rutherford of the University Library with thoroughness and care. Mr Francis Darwin has not only read the proofs but he willingly acceded to my request to write an Introduction supplementary to the account of his father's methods of work and treatment of books, with which readers of the

*Life and Letters* are familiar. I most cordially thank him for the confidence he has shown in a former pupil by placing the Library under his charge, and on behalf of the University I venture to assure him of the high appreciation of his generosity in rendering the Darwin Library accessible to all students.

The books are always available for reference on application ; it has, however, been deemed advisable to make a rule that they may not be removed from the Building.

A. C. SEWARD.

BOTANY SCHOOL, CAMBRIDGE,
*March 7th*, 1908.

# INTRODUCTION.

THE library of Charles Darwin has now found a permanent home in his own University, and it is perhaps appropriate that it should be in the Botany School, since it was a Cambridge professor of Botany who, more than any one man, determined his career as a naturalist.

The collection is not identical with that at Down. Thus the books he wrote and some few others from Down are in my own possession. There are also a few books of mine which, for the sake of convenience, are kept in the Darwin library: these are marked with an asterisk in the catalogue. Darwin's pamphlets are not included in the catalogue though part of them are on the shelves along side his books. The rest of the pamphlets are in the building and a manuscript catalogue of the whole is in my possession. His habit was to treat each pamphlet as a book, to number them and keep them in order on a shelf. But he was not consistent in the treatment of the publications received and many pamphlet-like volumes occur among his books.

He hardly ever had a book bound*, and the collection retains to a great degree its original ragged appearance. But some binding has necessarily been done, thus the copy of H. Müller's *Befruchtung* which he preserved "from complete dissolution by putting a metal clip over the back " † has now received more solid protection. The rough treatment he gave his books is described in the *Life and Letters*‡: an example may be seen in the sixth

---

* At one time he bound his pamphlets into volumes and two such volumes of *Philosophical Tracts* are preserved in his Library.

† *Life and Letters*, i. 150.                    ‡ i. 151.

edition of Lyell's *Elements* which he found too heavy to be read with comfort, and converted into two volumes by cutting in half.

The general characteristic of the library is incompleteness, hardly any sets of periodicals being perfect. The set of the Royal Society's publications is perhaps the best in this respect, and has a certain interest as representing several generations of fellowship of the Society.

The chief interest of the Darwin books lies in the pencil notes scribbled on their pages, or written on scraps of paper and pinned to the last page. Books are also to be found marked with a cypher, as described in *Life and Letters**.

Sir Joseph Hooker's essays on the Floras of Tasmania and New Zealand, which are bound together, are interesting as a memorial of the manner in which the two friends worked together. Thus near the end is a list in Hooker's handwriting of "Plants common to New Zealand and South America but not European," which is covered with pencilled notes by Darwin. This is followed by Darwin's abstract of the New Zealand essay annotated by Sir Joseph. In the middle of the volume is a list of New Zealand plants which was marked by Sir Joseph† to test Darwin's theory as to the sexes being commonly separated in trees‡.

Much of his reading was in German and a small memento of his difficulties with what he called the "verdammte" language§ occurs in a MS. translation of Rütimeyer's *Die Grenzen der Thierwelt*, made by the "patriotic German lady" over whom he used to triumph when she, naturally enough, hesitated in a *vivâ voce* translation of scientific German into English§. Darwin seems, in despair, to have given up Rütimeyer in the original, for all his notes are on the back of the translatress's manuscript.

Another book giving evidence of his manner of annotating as he read is Sprengel's *Das entdeckte Geheimniss der Natur*. Darwin has described‖ how he got this "wonderful book" in 1841 on the advice of Robert Brown, and how it stimulated the interest he

---

* i. 151.

† What must be an uncommon book was given to Darwin by Sir Joseph—viz. the privately printed *Notes on Norway* by his brother William Dawson Hooker.

‡ See *Origin of Species*. § *Life and Letters*, i. 126.

‖ *Life and Letters*, i. 90.

already felt in the fertilisation of flowers. Sprengel contains many interesting pencil notes by Darwin. Thus at p. 18 he wrote "seems to think fact of insects being required at all does not deserve an explanation, and how poor a one [i.e. an explanation] of Dicho-gamy,—for the convenience of insects!!" At p. 43 he has marked with triple pencil lines, as a sign of its importance, " Nature seems to have been unwilling that any flower should be fertilised by its own pollen." Hermann Müller* refers to this same passage in the *Entdeckte Geheimniss,* and points out how near Sprengel was to completely understanding the "secret of nature." Darwin had a personal feeling for the memory of Sprengel which comes out in the *Autobiography*† where he wrote:—"The merits of poor old Sprengel, so long overlooked, are now fully recognised many years after his death." Another book heavily annotated by Darwin is Gärtner's *Bastarderzeugung,* a work to which he often refers in his writings.

From a biographical point of view one of the most interesting books in the Library is a copy of Humboldt's *Personal Narrative.* Darwin wrote‡:—"This work...stirred up in me a burning zeal to add even the most humble contribution to the noble structure of Natural Science." He speaks of copying out from it "long pas-sages about Teneriffe" while he read aloud to his friends, who seem to have put him off with somewhat half-hearted promises of joining him in a natural history expedition.

And this book which so deeply affected him was given to him by Henslow, the master to whom he owed so much inspiration and to whom he was indebted for his appointment as naturalist in the *Beagle.* The inscription in vol. i. is "J. S. Henslow to his friend C. Darwin on his departure from England upon a voyage round the world. 21 Sepʳ 1831."

Another record of the voyage is his copy of Werner's *Nomen-clature of Colours* by "Patrick Syme, Flower-Painter, Edinburgh." It contains named samples of a number of different pigments, so that by matching a specimen with one of Patrick Syme's patches of colour it is possible to give a description which can be accurately interpreted. It was with this book that he described a cuttlefish at

---

* *Fertilisation of Flowers,* Eng. Trans., p. 3.          † *Life and Letters,* i. 91.
‡ *Life and Letters,* i. 55.

St Jago as "French grey" with bright yellow spots; and as its colour changed, he speaks of "clouds varying in tint between hyacinth red and chestnut brown" passing over its body\*. He recorded some observations in the volume itself: on a blank page at the beginning is written : " Beak of female ash grey, male nearly black, legs &c. exact dutch yellow."

In the course of preparing the catalogue we came across a publication by Darwin which had hitherto escaped notice. It is bound up in the second volume of *Philosophical Tracts* above referred to, and is entitled *A letter containing remarks on the moral state of Tahiti, New Zealand, &c.* by Capt. R. FitzRoy and C. Darwin Esq., of H.M.S. *Beagle*. It is dated "At sea, 28th June, 1836," is paged from 221 to 238, and headed "Read May 8, 1837"; where it was read and in what journal published I have not discovered. Darwin's contributions are marked D, and consist of extracts from his diary, many of which were ultimately published in the *Naturalist's Voyage*. Capt. FitzRoy in the character of commanding officer signs the paper. His signature is, however, followed by a paragraph signed by both authors.

" On the whole, balancing all that we have heard, and all that we ourselves have seen concerning the missionaries in the Pacific, we are very much satisfied that they thoroughly deserve the warmest support, not only of individuals, but of the British Government.

Robt. FitzRoy."<br>Charles Darwin."

In his Autobiography†, Darwin mentions his friendship with Dr Grant at Edinburgh University. Several of Grant's papers occur in the *Philosophical Tracts*, and one on the "Ciliæ" of the young of certain Mollusca has a minute point of interest, for the same unusual method of spelling "Cilia" occurs again in Darwin's *Naturalist's Voyage‡*. This heritage from Dr Grant would have made Dr Butler of Shrewsbury School turn in his grave.

The second volume of the *Philosophical Tracts* contains the "Extracts from letters addressed to Professor Henslow by C. Darwin,

---

\* *Naturalist's Voyage* (p. 7).      † *Life and Letters*, i. 38.
‡ Even in the recent editions.

Esq. Read at a meeting of the Cambridge Phil. Society, Nov. 16, 1835," and "printed for distribution among the Members" of the Society.

In his autobiography Darwin wrote*: "In October, 1838...I happened to read for amusement Malthus on Population†, and being well prepared to appreciate the struggle for existence which everywhere goes on...it at once struck me that under these circumstances favourable variations would tend to be preserved, and unfavourable ones to be destroyed."

The copy in the Library is disappointing, being inscribed " C. Darwin, 1841," and therefore not the volume he read in 1838. The first volume has been read and abstracted, but the second volume has not even been cut. A passage in vol. i., which is marked, is an opinion quoted from Dr Franklin: "were the face of the earth," he says, "vacant of other plants it might be gradually sowed and overspread with one kind only, as for instance with fennel." Why Dr Franklin chooses fennel for his mundane plant is not clear.

Another book connected with the *Origin of Species* is Patrick Matthew's *On Naval Timber and Arboriculture*, 1831. Darwin first heard of this book from the *Gardener's Chronicle* of Ap. 7, 1860‡, where Matthew published long extracts from it. Matthew claimed quite justly that he put forward the theory of Natural Selection long before the *Origin of Species* was published. It is certainly surprising to find in a book dated 1831, the expression "natural process of selection among plants."

Prichard's *Researches into the Physical History of Mankind* is interesting from another point of view. Darwin wrote in vol. i. (1851) "How like my Book this will be." Professor Seward and I have, I think, shown in *More Letters of Charles Darwin* that this refers to Prichard's work on geographical distribution and variation, not to his being an evolutionist. Yet Professor Poulton, using as it happens an edition of Prichard not in the possession of Darwin, came to the conclusion that his book contained a forecast of the theory of Descent.

The Library contains a good collection of books on various breeds of poultry, pigeons and other animals. Among these may

---

* *Life and Letters*, i. 83.   † An Essay on the Principle of Population, 1826.
‡ See *Life and Letters*, ii. 301, 302.

be mentioned Eaton's *Treatise on the Almond Tumbler*, 1851, re-printed in both editions of his *Treatise on tame, domesticated and fancy pigeons*. Mr Eaton doubtless belonged to the Columbarian and Philoperistera Clubs\*, of which Darwin was, I think, a member. He wrote (*loc. cit.* p. 281), "I have found it very important asso-ciating with fanciers and breeders....I sat one evening in a gin palace in the Borough among a set of pigeon fanciers, when it was hinted that Mr Bull had crossed his Pouters with Runts to gain size." He describes the fanciers shaking their heads over this "scandalous proceeding," and goes on to draw a moral as to the bearing of crossing on fancying. He was fond of quoting the following passage which occurs at p. vi of the introduction to the *Treatise on the Almond Tumbler†*:

"If it was possible for noblemen and gentlemen to know the amazing amount of solace and pleasure derived from the Almond Tumbler, when they begin to understand their [the tumbler's] properties, I should think that scarce any nobleman or gentleman would be without their aviaries of Almond Tumblers."

Huth's *Marriage of Near Kin*, 1875, is mentioned in Darwin's autobiography‡ as containing one of the "three intentionally falsi-fied statements" known to him. Mr Huth (pp. 298, 302) quoted the work of a Belgian on rabbit breeding; Darwin instinctively doubted the genuineness of the account which turned out to be a fraud. Mr Huth did what he could to correct the mistake into which he had been led, by inserting a slip in the unsold copies of his book.

Ziegler's *Atonicité et Zoïcité* contains inaccuracies of a different kind. The author doubtless believed that he saw the extraordinary reactions with *Drosera* which Darwin was unable to confirm. The passage in *Insectivorous Plants* (p. 21) is a good instance of Darwin's respectful treatment of inexperienced observers. Another book containing statements not generally received is Otto Hahn's *Die Urzelle*, in which the fossil remains of plants

* *Life and Letters*, ii. 51.

† Quoted in *Life and Letters*, ii. 52.

‡ *Life and Letters*, i. 106. The title is here incorrectly given as "Consanguineous Marriage."

are described and figured as occurring in granite and similar improbable localities.

Andrew Smith's *Illustrations of the Zoology of South Africa* (1849) is connected in my mind with early recollections of my father. Looking through the coloured plates was converted into an exciting game by the supposition that the birds belonged alternately to himself and to the child who was his playfellow. I can well remember (after a series of dull thrush-like birds had been calmly shared between my father and myself) the agony of seeing a magnificent green and purple one fall to his lot. I am sure he tried to cheat himself, but this was not always possible.

In the foregoing pages I have not attempted to do more than call attention to a few points of general interest in relation to the Darwin Library. I have not the knowledge to answer the question that will naturally suggest itself,—what rare books does the collection contain? I can only say that a casual comparison of parts of the catalogue with that of the University Library shows that we have a certain number of books not in that collection.

Finally it is a pleasure to express my appreciation of Professor Seward's kindness in finding room for my books in the Botany School. It is a satisfaction to know that they will be well cared for, not only by their present guardian (and long may he remain in charge of them), but also by future occupants of the Chair of Botany in Cambridge.

FRANCIS DARWIN.

Botany School, Cambridge.
*March* 4, 1908.

## ABBREVIATIONS AND SYMBOLS.

\*    Books thus marked were not in the Library at Down.

[  ]    Indicates a name or a title not occurring on the title page: *e.g.* [Philosophical Tracts], a title taken from the back of a bound volume of pamphlets.

Extr.    Indicates a reprint from a journal etc.

n. d.    An undated book.

Pr. pr.    Privately printed.

Pseud.    Means that the name given in the Catalogue is a pseudonym.

---

## ERRATUM.

Delamer (E. S.) should have been catalogued as a pseudonym for Edmund S. Dixon.

# CATALOGUE OF THE DARWIN LIBRARY.

**Abercrombie** (John). Inquiries concerning the Intellectual Powers and the Investigation of Truth. 8th ed. 8vo. *London*, 1838. **12**

**\*Abernethy** (John). Physiological Lectures. 2nd ed. 8vo. *London*, 1822. **115**

**Acébla** (Alexandre d'). Les Impiétés. 4to. *Paris*, 1878. **Ne**

**Acharius** (Erik). Methodus qua omnes detectos lichenes, &c. Sectio 1, 2. 8vo. *Stockholmiae*, 1803. **62**

**Acta** (Nova) reg. soc. scientiarum Upsaliensis. Ser. 3, Vols. 1–3, 6–10. 4to. 1855–79. **44**
In Memoriam Quattuor Seculorum ab Universitate Upsaliensi peract-orum. 4to. 1877. **44**

**Actes** du Congrès de **botanique** horticole réuni à Bruxelles...le 1er mai 1876. Rédigés par Edouard Morren. 8vo. *Liége*, 1877. **59**

**Adams** (A. Leith). Field and Forest Rambles, &c. 8vo. *London*, 1873. **26**

**Adams** (John). *See* Juan (George). **95**

**Adriasola y Arve** (J. M.). Reflexiones Medicas sobre el Analisis de las Aguas de Yura. (Extr.) 8vo. n. d. [Philos. Tracts, i. 23.] **11**

**Agassiz** (Alex.). North American Acalephæ. (Extr.) 4to. *Cambridge, Mass.*, 1865. **74**
Revision of the Echini. 4 Parts and Plates. (Extr.) 4to. *Cambridge, Mass.*, 1872–74. **74**
Echini, Crinoids, and Corals. By A. A., and L. F. de Pourtalès. (Extr.) 4to. *Cambridge, Mass.*, 1874. **74**
North American Starfishes. (Extr.) 4to. *Cambridge, Mass.*, 1877. **74**
Report on the Echinoidea (The Zoology of the Voyage of H.M.S. Challenger...1873–76. Vol. 3, Part 9). 4to. *London*, 1881. **67**

**Agassiz** (Elizabeth C.). Seaside Studies in Natural History. By E. C. A., and Alex. Agassiz. 8vo. *Boston*, 1871. **26**

**Agassiz** (Louis). Nomenclatoris Zoologici Index Universalis. 8vo.
*Soloduri*, 1848. 106
Principles of Zoölogy. Part 1. Comparative Physiology. By L. A.,
and A. A. Gould. 8vo. *Boston*, 1848. 106
Lake Superior...with a narrative of the Tour, by J. Elliot Cabot. 8vo.
*Boston*, 1850. 15
Bibliographia Zoologiæ et Geologiæ. (Ray Soc. Publ.) Vol. 3, ed. by
H. E. Strickland; Vol. 4, ed. by H. E. S., and Sir W. Jardine. 8vo.
*London*, 1852–54. 17
Contributions to the Natural History of the United States of North
America. Vol. 1, Part 1, 232 pp. 4to. 8
Methods of Study in Natural History. 8vo. *Boston*, 1863. 26
Address delivered on the Centennial Anniversary of the birth of
Alexander von Humboldt. 8vo. *Boston*, 1869. 113
De l'Espèce et de la classification en Zoologie. Trad. de l'anglais par
F. Vogeli. Éd. revue...par l'auteur. 8vo. *Paris*, 1869. 112
Report on the Florida Reefs. (Extr.) 4to. *Cambridge, Mass.*, 1880. 74
*See* Morton (S. G.). 114

**\*Ainsworth's** Latin Dictionary. 67

**Alder** (Joshua). A Monograph of the British Nudibranchiate Mollusca.
By J. A., and Albany Hancock. Parts 1–7. (Ray Soc. Publ.)
Fol. *London*, 1845–55. 72

**Alglave** (Émile). *See* Bernard (C.). 115

**Allan** (Robert). *See* Phillips (Wm.). An elementary introduction to
Mineralogy. 8vo. *London*, 3rd ed., 1823; 4th ed., 1837. 117

**Allen** (Grant). Physiological Æsthetics. 8vo. *London*, 1877. 12
The Colour-Sense : its origin and development. 8vo. *London*, 1879. 12
Der Farbensinn. Deutsche Ausg....von Ernst Krause. (Darwinistische
Schriften, Nr. 7.) 8vo. *Leipzig*, 1880. 41

**Allen** (Joel Asaph). The American Bisons, living and extinct. (Extr.)
4to. *Cambridge, Mass.*, 1876. 74
History of North American Pinnipeds. 8vo. *Washington*, 1880. 118

**Allman** (George J.). A Monograph of the Fresh-Water Polyzoa. (Ray
Soc. Publ.) Fol. *London*, 1856. 72
A Monograph of the Gymnoblastic or Tubularian Hydroids. In 2 Parts.
(Ray Soc. Publ.) Fol. *London*, 1871–72. 72
Report of the Hydroida collected...by L. F. de Pourtalès. 4to. *Cam-
bridge, Mass.*, 1877. 74

**Altum** (B.). Zoologie. Von B. Altum und H. Landois. 2te Aufl.
8vo. *Freiburg i. B.*, 1872. 106

**American Academy** of Arts and Sciences. Proceedings. 8vo. *Boston*.
Vols. 1–3, 1848–. Vols. 5–7 (incomplete), 1860–. N. S. Vols. 1–6,
8 = whole ser., Vols. 9–14, 16, 1874–. Ng

**American Association** for the Advancement of Science. Proceedings.
2nd Meeting, August, 1849. 8vo. *Boston*, 1850. 119

# 3

**American Journal** of Science and Arts. Vol. 38, Nos. 1, 2, 1840; Vol. 45, No. 2, 1843; 3rd Ser., Vol. 11, Feb. 1876. **116**

**American Naturalist.** Vols. 5–17 (incomplete). 8vo. *Salem, Mass.*, 1871–83. **19**

**American Philosophical Society.** Proceedings. Nos. 83–89, 92–101, 107–109 (Vols. XI.–XIX.). 8vo. 1870–. **116**

**Anales** del Museo Público de Buenos Aires. Por G. Burmeister. Tom. 1, 2. 4to. 1864–74. **72**

**Anderson** (John). A Report on the Expedition to Western Yunan *viâ* Bhamô. 8vo. *Calcutta*, 1871. **75**

**Andres** (Angelo). Le Attinie. Vol. 1°. (Fauna und Flora des Golfes von Neapel, ix. 1.) 4to. *Leipzig*, 1884. **72**

**Angelin** (N. P.). Iconographia Crinoideorum in stratis Sueciæ siluricis fossilium....Opus postumum edendum curavit Regia Academia Scientiarum Suecica. Fol. *Holmiæ*, 1878. **Q. 1**

**Annales des sciences naturelles.** 2ᵉ série, Tomes 15, 16. Jan.–Mai, Juillet–Déc., 1841. 8vo. *Paris.* **35**

**Annals and Magazine** of Natural History. Vols. 6–20 (1st ser.). Vols. 1, 2, 13–16 (2nd ser.). Vols. 15, 16 (3rd ser.). Vol. 17–20 (3rd ser.). 4th ser. Vols. 1–12. 8vo. London, 1841–73. **43a & 37**

**Annals of Natural History.** Conducted by Sir W. Jardine, Bart. (and others). Vols. 1, 2, 1838–39. 8vo. *London.* **43a**

**Anthropologia.** Vol. 1, Nos. 1, 2. 8vo. *London*, 1873–74. **42**

**Anthropological Institute** of Great Britain and Ireland. Journal. Vols. 1–4. 8vo. *London*, 1872–75. **42**

**Anthropological Review** and Journal of the Anthropological Society of London. Nos. 1–13, 18–27. 8vo. *London*, 1863–69. **42**

**Anthropological Society** of London. Journal. *See* Anthropological Review, 1863–. **42**
Memoirs. Vols. 1, 3. 8vo. *London*, 1863–69. **42**

**Archiac** (*Le Vicomte* d'). Histoire des progrès de la Géologie de 1834 à 1845. Tome 1er. 8vo. *Paris*, 1847. **107**

**Archiv** für **Anthropologie.** 1er Bd, 2tes Heft; 3er Bd, 3tes u. 4tes Heft. 4to. *Braunschweig*, 1866, 1869. **Ne**

**Areschoug** (F. W. C.). Beiträge zur Biologie der Holzgewächse. (Extr.) 4to. *Lund*, 1877. **44**

**Argyll** (The Duke of). The Reign of Law. 8vo. *London*, 1867. **22**
Primeval Man. 8vo. *London*, 1869. **114**

**Aristotle** on the parts of Animals. Transl. with introd. and notes, by W. Ogle. 8vo. *London*, 1882. **Nf**

1—2

4

**Arnott** (George A. W.). *See* Hooker (*Sir* W. J.). The British Flora. 7th ed. 8vo. *London*, 1855. 59

**Arnott** (Neil). Elements of Physics. Vol. 1; Vol. 2, Part 1, 5th ed. 8vo. *London*, 1833. 105

**Asiatic Society** of Bengal. Journal. Vol. 40, Nos. 169–171; Vol. 41, No. 174; N. S. Vol. 43, Part 2, Extra No.; Vol. 49. 8vo. *Calcutta*, 1871–80. 119
Proceedings. Sept.–Nov. 1871; Feb. 1882. 8vo. *Calcutta*. 119

**Askenasy** (E.). Beiträge zur Kritik der Darwin'schen Lehre. 8vo. *Leipzig*, 1872. 39

**Atkinson** (E.). *See* Helmholtz (H.). Popular Lectures on Scientific Subjects. Transl. 8vo. *London*, 1873. 10

***Atlas** (Astronomy). *See* Milner (*Rev.* T.). Nc
*(England and Wales). British Atlas. By J. and C. Walker. Fol. *London*, 1837. Q. 1
(Publ. by the Society for the Diffusion of Useful Knowledge, 1844–.) Incomplete. Fol. *London*. C. Knight & Co. Q. 3
*A new general Atlas. By Sidney Hall. New ed. Fol. *London*, n. d. Q. 1
do. Alphabetical Index of all the names. 8vo. *London*, 1831. Nd
(Zoological). *See* McAlpine (D.). Q. 1
*(de la Géographie). Atlas complet du Précis de la Géographie universelle de Malte-Brun...revu et corrigé par J. J. N. Huot. Fol. *Paris*, 1837. Q. 1
*der Pflanzenverbreitung. *See* Drude (Oscar). Fol. *Gotha*, 1887. Q. 1

**Atti d. r. accad. dei Lincei.** Ser. 2ᵃ, Vol. 3, Fasc. 1°, 2°; Vols. 5, 7. 4to. *Roma*, 1876, 1880. 42
Ser. 3ᵃ. Memorie della classe di scienze morali, storiche e filologiche. Vol. 3. 4to. *Roma*, 1879. 42
Ser. 3ᵃ. Memorie della classe di scienze fisiche, matematiche e naturali. Vols. 3, 7, 8. 4to. *Roma*, 1879–80. 42
*dell' istituto botanico dell' università di **Pavia.** Redatti da G. Briosi. II. Ser., Vol. 7. 8vo. *Milano*, 1902. Nb
d. r. accad. delle scienze di **Torino.** Classe di scienze fisiche e matematiche. Vol. 14, 1, 4–7. 8vo. 1878–79. 18

**Aubuisson de Voisins** (J. F. d'). Traité de Géognosie. Tome 1, 2. 8vo. *Strasbourg*, 1819. 117

**Audubon** (John J.). Ornithological Biography: Birds of the United States of America. 5 vols. 8vo. *Edinburgh*, 1831–39. 75
The Viviparous Quadrupeds of North America. By J. J. A., and the Rev. J. Bachman. Vol. 1. 8vo. *New York*, 1846. 75

**Auinger** (M.). *See* Hoernes (R.). 72 & Nb

**\*Authors and Publishers.** A Manual of Suggestions for Beginners in Literature. By G. H. P. and J. B. P. 7th ed. 8vo. *New York,* 1897. **24**

**Avebury** (*Lord*). *See* Lubbock (John).

**Aveling** (Edward B.). The Student's Darwin. 8vo. *London,* 1881. **23**
\*Die Darwin'sche Theorie. 8vo. *Stuttgart,* 1887. **40**

**Aviary** (The). *See* British Aviary. **127**

**Ayrault** (Eugène). De l'industrie mulassière en Poitou. 8vo. *Niort,* 1867. **11**

**Azara** (Félix d'). Essais sur l'histoire naturelle des Quadrupèdes de la Province du Paraguay. 2 tomes. 8vo. *Paris,* An IX (1801). **108**
Voyages dans l'Amérique méridionale...1781-1801. Éd. par C. A. Walckenaer. Accomp. d'un Atlas de 25 planches. 4 tomes. 8vo. *Paris,* 1809. **15 & 72**

**Babington** (Charles C.). Manual of British Botany. 3rd ed. 8vo. *London,* 1851. **59**

**Baerenbach** (Friedrich von). Das Problem einer Naturgeschichte des Weibes. 8vo. *Jena,* 1877. **40**
Gedanken ueber die Teleologie in der Natur. 8vo. *Berlin,* 1878. **40**
Prolegomena zu einer Anthropologischen Philosophie. 8vo. *Leipzig,* 1879. **114**

**Bagehot** (Walter). Physics and Politics. 8vo. *London,* 1872. **11**

**Baildon** (Henry Bellyse). The Spirit of Nature. 8vo. *London,* 1880. **23**

**Bain** (Alexander). The Senses and the Intellect. 2nd ed. 8vo. *London,* 1864. **22**
The Emotions and the Will. 2nd ed. 8vo. *London,* 1865. **12**

**Bain** (Andrew Geddes). On the discovery of the Fossil Remains of *Bidental*...in South Africa. (Extr.) (*See* R. Owen.) 4to. *London,* 1845. **Nc**

**Baker** (J. G.). Elementary Lessons in Botanical Geography. 8vo. *London,* 1875. **62**

**Balfour** (Francis M.). *See* Foster (M.). The Elements of Embryology. 8vo. *London,* 1874. **106**
A Monograph on the development of Elasmobranch Fishes. 8vo. *London,* 1878. **106**
A treatise on Comparative Embryology. 2 vols. 8vo. *London,* 1880-81. **39**
\*The Works of F. M. Balfour. Ed. by M. Foster and Adam Sedgwick. 4 vols. Memorial Edition. 8vo. *London,* 1885. **Nd**

**Ball** (John). *See* Hooker (J. D.). Journal of a Tour in Marocco, &c. 8vo. *London,* 1878. **89**

**Ball** (V.). Jungle life in India. 8vo. *London,* 1880. **9**

**Ball** (Wm. Platt). Are the effects of use and disuse inherited? 8vo. *London*, 1890. 11

**Baltzer** (Joh. Bapt.). Ueber die Anfänge der Organismen, &c. 3te Aufl. 8vo. *Paderborn*, 1870. 9

**Baly** (Wm.). *See* Müller (J.). Elements of Physiology. Transl. 2 vols. 8vo. *London*, 1838-42. 125
Recent advances in the Physiology of Motion, the Senses, &c. By W. B., and W. Senhouse Kirkes. 8vo. *London*, 1848. 125

**Barclay** (John). An Inquiry into the Opinions ancient and modern concerning Life and Organization. 8vo. *Edinburgh*, 1822. 39

**Barker** (Arthur E. J.). *See* Frey (H.). The Histology and Histo-chemistry of Man. Transl. 8vo. *London*, 1874. 115

**Barker-Webb** (P.). Histoire naturelle des Îles Canaries. Par P. B.-W., et S. Berthelot. Tome 3ème, 1ère partie. Géographie botanique. 4to. *Paris*, 1840. **Nb**

**Barrago** (Francesco). L' Uomo fatto ad imagine di Dio fu anche fatto ad imagine della scimia. 8vo. *Cagliari*, 1869. 39

**Barrande** (Joachim). Défense des Colonies. IV. 8vo. *Prague*, 1870. 14
Céphalopodes. Extr. du syst. silurien du centre de la Bohême. 8vo. *Prague*, 1870 & 1877. 117
Trilobites. Extr. du syst. silurien du centre de la Bohême. 8vo. *Prague*, 1871. 117
Brachiopodes. Extr. du syst. silurien du centre de la Bohême. 8vo. *Prague*, 1879. 117
Acéphalés. Extr. du syst. silurien du centre de la Bohême. 8vo. *Prague*, 1881. 117·

**Barrow** (*Sir* John), *Bart.* Sketch of the surveying voyages of H.M.Ss. Adventure and Beagle, 1825-36. (Extr.) 8vo. *London*.
[Philos. Tracts, ii. 1.] 11

**Barton** (John). A Lecture on the Geography of Plants. 12mo. *London*, 1827. 62

**Bary** (A. de). Die Mycetozoen. (Schleimpilze.) 2te Aufl. 8vo. *Leipzig*, 1864. 55

**Bastian** (H. Charlton). The modes of origin of Lowest Organisms. 8vo. *London*, 1871. 102
The Beginnings of Life. 2 vols. 8vo. *London*, 1872. 9
Evolution and the Origin of Life. 8vo. *London*, 1874. 9
The Brain as an Organ of Mind. 8vo. *London*, 1880. 11

**Bate** (C. Spence). *See* British Museum. Catalogue of the specimens of Amphipodous Crustacea. 8vo. *London*, 1862. 102

**Bateman** (Frederic). On Aphasia, or Loss of Speech. 8vo. *London*, 1870. 92

**Bates** (Henry W.). The Naturalist on the River Amazons. 2 vols. 8vo. *London*, 1863. 89

**\*Bateson** (William). Materials for the study of Variation. 8vo. *London*, 1894. 39

**\*Bauhin** (Johannes). Historia plantarum universalis. Auctoribus J. B., J. H. Cherlero. Quam recensuit et auxit Dominicus Chabræus. Tom. 3. Fol. *Ebroduni*, 1651. 72

**Baxter** (J. H.). Statistics, medical and anthropological...derived from records of the Examination for Military Service, &c. 2 vols. 4to. *Washington*, 1875. 74

**Bayfield** (*Capt.* H. W.), *R.N.* Notes on the Geology of the North Coast of the St. Lawrence. (Extr.) 4to. *London* [Read 1833]. **Nc**

**Beale** (Lionel S.). On the structure and growth of the Tissues, and on Life. 8vo. *London*, 1865. 115

**Beaumont** (Élie de). Note sur les systèmes de Montagnes les plus anciens de l'Europe. (Extr.) 8vo. *Paris*, 1847. 117
Leçons de Géologie pratique. Tome 1er. 8vo. *Paris*, 1845. 107
*See* Dufrénoy. 107

**Bechstein** (J. M.). Gemeinnützige Naturgeschichte Deutschlands nach allen drey Reichen. 4 Bde. 8vo. *Leipzig*, 1793– [1te Aufl. = Bde 3, 4; 2te Aufl. = Bde 1, 2]. 8
Naturgeschichte der Stubenvögel. 4te Aufl. 8vo. *Halle*, 1840. 127

**Beechey** (*Admiral* F. W.), *R.N.* Narrative of a voyage to the Pacific and Beering's Strait...1825–28. 8vo. *Philadelphia*, 1832. 16

**Bell** (Alexander Graham). Memoir upon the formation of a deaf variety of the human race. (Extr.) 4to. *New Haven*, 1883. 72

**\*Bell** (Arthur J.). Why does man exist? 8vo. *London*, 1890. 26

**Bell** (*Sir* Charles). The anatomy and philosophy of Expression as connected with the Fine Arts. 3rd ed. 8vo. *London*, 1844. **Nf**
*The Hand. Edition 9. *London*, 1874. 24

**Bell** (F. Jeffrey). *See* Gegenbaur (Carl). 92

**Bell** (John and Charles). The Anatomy and Physiology of the Human Body. 6th ed. 3 vols. 8vo. *London*, 1826. 94

**Belt** (Thomas). The Naturalist in Nicaragua. 8vo. *London*, 1874. 9

**Beneden** (P. J. van). Mémoire sur les Vers intestinaux. (Suppl. aux Comptes rendus, Tome 2.) 4to. *Paris*, 1861. **Ne**

**Benza** (P. M.). Notes, chiefly geological, of a Journey through the Northern Circars in the year 1835. (Extr.) 8vo. [Philos. Tracts, ii. 15.] 11
Notes on the Geology of the country between Madras and the Neilgherry Hills, &c. 8vo. (1836). [Philos. Tracts, ii. 14.] 11

\*Berghaus' Physikalischer Atlas, Abt. v. *See* Drude (Oscar). Fol. *Gotha*, 1887. Q. 1

Bericht der 50. Versammlung deutscher Naturforscher und Aerzte in München vom 17. bis 22. September, 1877. 4to. 75

Berjeau (Ph. Charles). The varieties of Dogs, as they are found in old Sculptures, &c. 4to. *London*, 1863. 118

Bernard (Claude). Leçons sur les propriétés des Tissus vivants. Recueillies, rédigées et publ. par Émile Alglave. (Extr.) 8vo. *Paris*, 1866. 115
Leçons sur les phénomènes de la vie communs aux animaux et aux végétaux. Tome 2ème. 8vo. *Paris*, 1879. 115

Bernhardi (Johann Jacob). Ueber den Begriff der Pflanzenart und seine Anwendung. 4to. *Erfurt*, 1834. 60

Berthelot (Sabin). *See* Barker-Webb (P.). Nb

Berzelius (J. J.). The use of the Blowpipe in chemical analysis, &c. By J. J. B. Transl. from the French of M. Fresnel, by J. G. Children. 8vo. *London*, 1822. 117

Besant (Annie). *See* Büchner (L.). Mind in Animals. Transl. 8vo. *London*, 1880. 118

Bethune (C. R. Drinkwater). *See* Hawkins (*Sir* R.). 89

Beudant (F. S.). Traité élémentaire de Minéralogie. 2ème éd. Tome 1er. 8vo. *Paris*, 1830. 117

Bevan (Edward). The Honey-Bee. 8vo. *London*, 1827. 95

Bevan (Hugh J. C.). *See* Pouchet (G.). The Plurality of the Human Race. Transl. 8vo. *London*, 1864. 12

Bevington (L. S.). Key-Notes. 8vo. *London*, 1879. 24

Bianconi (J. Joseph). La Théorie Darwinienne et la Création dite indépendante. Lettre à M. Ch. Darwin. 8vo. *Bologne*, 1874. 13
La Teoria Darwiniana e la Creazione detta indipendente...Lettera al Signor Carlo Darwin. Trad. dal Francese da G. A. Bianconi. Riveduta ed accres. 8vo. *Bologna*, 1875. 13

Bienen-Zeitung. Organ des Vereins deutscher Bienenwirthe. 12ter Band. Jahrg. 1856. 4to. *Nördlingen*. Ne

Bigg (R. Heather). Spinal Curvature. 8vo. *London*, 1882. 94

Bikkers (Alex. V. W.). *See* Schleicher (A.). Darwinism tested by the science of Language. Transl. 8vo. *London*, 1869. 10

Billing (Sidney). Scientific materialism and ultimate conceptions. 8vo. *London*, 1879. 28

**Binney** (W. G.). The terrestrial air-breathing Mollusks of the United States, &c. With Plates. ["Terrestrial Mollusks," Vol. 5.] 2 vols. 8vo. *Cambridge, Mass.*, 1878. **Nf**

**Birds.** (The Naturalist's Library, Ed. by *Sir* W. Jardine, *Bart.*, Vols. ix and xiv.) 8vo. *Edinburgh.* **118**

**Bischoff** (Ludwig W. T. v.). *See* Lucae (J. C. G.). **72**

**Black** (John). *See* Buch (L. von). **Ne**
*See* Humboldt (Alex. von). Political Essay on the Kingdom of New Spain. Transl. 2 vols. 8vo. *New York*, 1811. **16**

**Blackley** (Charles H.). Experimental researches on...Catarrhus aestivus. 8vo. *London*, 1873. **104**
Hay Fever. 2nd ed. 8vo. *London*, 1880. **104**

**Blackwall** (Antoinette B.). Studies in general Science. 8vo. *New York*, 1869. **26**

**Blackwall** (John). Researches in Zoology. 8vo. *London*, 1834. **112**
A History of the Spiders of Great Britain and Ireland. 2 Parts. (Ray Soc. Publ.) Fol. *London*, 1861–64. **72**

**Blainville** (D. de). *See* Dictionnaire des Sciences naturelles. Planches. Zoologie. 8vo. *Paris*, 1816–30. **41**

**Blainville** (H. M. D. de). Manuel d'Actinologie ou de Zoophytologie. Avec un Atlas. 2 vols. 8vo. *Paris*, 1834. **96**

**Blake** (C. Carter). *See* Broca (P.). On the Phenomena of Hybridity in the Genus Homo. 8vo. *London*, 1864. **124**

**Blanford** (W. T.). *See* Medlicott (H. B.) **Nf**

**\*Blomefield** (Leonard), (late Jenyns). Chapters in my Life. (Pr. pr.) Reprint. 8vo. *Bath*, 1889. **113**
*See* Jenyns (L.).

**Blumenbach** (Johann F.). The Anthropological Treatises of J. F. B.... Transl. and ed. by Th. Bendyshe. 8vo. *London*, 1865. **114**

**Blyth** (Edward). The natural history of the Cranes...enlarged, and repr....by W. B. Tegetmeier. 8vo. *London*, 1881. **Nf**

**Boitard.** Les Pigeons de volière et de colombier. Par MM. Boitard et Corbié. 8vo. *Paris*, 1824. **127**
Manuel d'Entomologie. 2 tomes. 12mo. *Paris*, 1828. **118**

**Bollettino** meteorologico ed astronomico del r. Osservatorio dell' Università di Torino. Anno vii., 1872. **8**

**Bon (Le) Jardinier,** pour l'année 1827. 28ème éd. 12mo. *Paris*, 1827. **24**

**Bonaparte** (*Prince* Ch. I..). A geographical and comparative list of the Birds of Europe and North America. 8vo. *London*, 1838.    127
The State of Zoology in Europe. Transl. by H. E. Strickland. (Ray Soc. Publ. Reports, 1841, 1842.) 8vo. *London*, 1845.    17
Coup d'œil sur l'ordre des Pigeons. (Extr.) 4to. *Paris*, 1855.    **Ne**

**Bondi** (Augusto). L' Uomo. Ipotesi sulla Origine (Teoria Darwiniana), considerazioni. 8vo. *Forlì*, 1873.    9

**\*Boner** (Charles). Transylvania; its products and its people. 8vo. *London*, 1865.    25

**Bonnal** (Marcel de). Une Agonie. 8vo. *Angoulême*, 1877.    28

**\*Bonnet** (Charles). Recherches sur l'usage des feuilles dans les Plantes. 4to. *A Gottingue*, 1754.    61
Œuvres d'Histoire naturelle et de Philosophie. Tomes 1, 2. Insectologie. 8vo. *Amsterdam*, 1780.    95

**Boott** (Francis). Illustrations of the Genus Carex. 3 Parts. Fol. *London*, 1858–62.    **Q. 1**

**Borrelli** (Diodato). Vita e Natura. 8vo. *Napoli*, 1879.    12

**Bosquet** (J.). Description des Entomostracés fossiles des terrains tertiaires de la France et de la Belgique. (Extr.) 4to. *Bruxelles*, 1852.    **Ne**
Monographie des Crustacés fossiles du terrain crétacé du Duché de Limbourg. 4to. *Haarlem*, 1854.    **Na**
Notice sur quelques Cirripèdes récemment découverts dans le terrain crétacé du Duché de Limbourg. 4to. *Haarlem*, 1857.    74
Monographie des Brachiopodes fossiles du terrain crétacé supérieur du Duché de Limbourg. 1ère Partie. (Extr.) 4to. *Haarlem*, 1859.    **Na**

**\*Bostock** (John). An elementary system of Physiology. Vol. 1. 8vo. *London*, 1824.    115

**Boston** Society of Natural History. Memoirs II. i. 1–3; II. ii. 1–4; II. iii. 1–5; II. iv. 1–6; III. i. 1–3. 4to. *Boston*, 1869–79.    **Ne**
Occasional Papers. Nos. 2, 3. 1875–80.    104
Proceedings. Vols. 13–19; 20, Parts 2, 3. 1869–80.    104

**Botanical Magazine;** or, Flower-Garden Displayed. *See* Curtis (Wm).    59

**Bouchard** (A.). *See* Wundt (W.). Nouveaux éléments de Physiologie humaine. Trad. 8vo. *Paris*, 1872.    115

**Boudin** (J. Ch. M.). Traité de géographie et statistique médicales et des maladies endémiques. 2 tomes. 8vo. *Paris*, 1857.    94

**Boué** (Ami). Autobiographie du Docteur médécin Ami Boué...1794. 8vo. *Vienne*, 1879.    113
Catalogue des Œuvres...du Dr Ami Boué. 8vo. *Vienne*, 1876. [Published with the " Autobiographie."]    113

**Bourbon del Monte** (*Marquis* J.-B. François). L'homme et les animaux. Essai de Psychologie positive. 8vo. *Paris*, 1877.   47

**Bouverie-Pusey** (S. E. B.). Permanence and Evolution. 8vo. *London*, 1882.   40

**\*Bower** (F. O.). Practical Botany for Beginners. By F. O. B., and D. T. Gwynne-Vaughan. 8vo. *London*, 1902.   24

**Bowerbank** (J. S.). A Monograph of the British Spongiadæ.  4 vols. [Vol. 4, ed. by Rev. A. M. Norman] (Ray Soc. Publ.). 8vo. *London*, 1864–72.   17

**\*Boyer** (A.). *See* Dictionary (French).   67

**Brace** (Charles L.). The Races of the Old World. 8vo. *London*, 1863.
114
The dangerous classes of New York. 8vo. *New York*, 1872.   28

**Bradley** (Richard). A general treatise of Husbandry and Gardening. 3 vols. 8vo. *London*, 1724.   50

**Brady** (G. Stewardson). A monograph of the...Copepoda of the British Islands. 3 vols. (Ray Soc. Publ.) 8vo. *London*, 1878–80.   17a

**Braun** (A.). *See* Henfrey (A.). Botanical and Physiological Memoirs. Transl. 8vo. *London*, 1853.   17

**Bree** (C. R.). Species not Transmutable. 8vo. *London* (1860).   10

**Brehm** (A. E.). Illustrirtes Thierleben. 4 Bde. 8vo. *Hildburghausen*, 1864–67.   Nd
do.  Grosse Ausgabe. 2te umgearb. Aufl. 1te–4te Abt. 8vo. *Leipzig*, 1876–78.   Nd

**Brent** (B. P.). The Canary...and some other Birds. 8vo. *London*, n. d.
128
The Pigeon Book. 8vo. *London*, n. d.   117

**Briggs** (T. R. A.). Flora of Plymouth. 8vo. *London*, 1880.   62

**Briosi** (Giovanni). Intorno un organo di alcuni embrioni vegetali. (Extr.) 4to. *Roma*, 1882.   Nb
*See* Atti dell' ist. bot. dell' Università di Pavia. II. Ser., Vol. 7. 8vo. *Milano*, 1902.   Nb

**British Association** for the Advancement of Science. Report of the 3rd Meeting...Cambridge, 1833. 8vo. *London*, 1834.   16
Report of the 11th Meeting...Plymouth, 1841. 8vo. *London*, 1842.   16

**British** (The) Aviary. 8vo. *London*, n. d.   127

**British Museum.** Catalogue [List] of the specimens of Mammalia... (Ed. by J. E. Gray.) Parts 1–3. 12mo. *London*, 1843-52.   96
Catalogue of Marine Polyzoa...Parts 1, 2. (Prepared by G. Busk. Ed. by J. E. Gray.) 8vo. *London*, 1852–54.   96

Catalogue of British Hymenoptera...By Fr. Smith. Part 1. Apidæ—
Bees. 8vo. *London*, 1855. 96
Catalogue of the Coleopterous Insects of Madeira...By T. Vernon
Wollaston. 8vo. *London*, 1857. 102
Catalogue of the specimens of Amphipodous Crustacea...By C. Spence
Bate. 8vo. *London*, 1862. 102
Catalogue of the Chiroptera...By G. E. Dobson. 8vo. *London*, 1878.
102

**\*British Museum and Kew.** Committee on Botanical Work. Report
...dated 11th March, 1901. Fol. *London.* Nc
\*Minutes of Evidence...to accompany the Report...dated 11th March,
1901. Fol. *London.* Nc

**Broca** (Paul). On the Phenomena of Hybridity in the Genus Homo.
Ed. by C. Carter Blake. 8vo. *London*, 1864. 124

**Brodie** (*Sir* Benj. C.), *Bart. See* Psychological Inquiries. 8vo. *London*,
1854. 12

**Brongniart** (Alexandre). *See* Dictionnaire des Sciences naturelles.
Planches. Minéralogie. 8vo. *Paris*, 1816–30. 41

**Bronn** (Heinrich G.). Essai d'une réponse à la question de prix proposée
en 1850 par l'Académie des Sciences pour le concours de 1853...
savoir : Étudier les lois de la distribution des corps organisés fossiles,
&c. (Suppl. aux Comptes rendus, Tome 2.) 4to. *Paris*, 1861. Ne
Handbuch einer Geschichte der Natur. 2 Bde. 8vo. *Stuttgart*, 1842–43.
26
Morphologische Studien über die Gestaltungs-Gesetze der Naturkörper,
&c. 8vo. *Leipzig*, 1858. 106
Untersuchungen über die Entwickelungs-Gesetze der organischen Welt.
Deutsch hrsg. von H. G. B. 8vo. *Stuttgart*, 1858. 107

**Brookes** (R.). The Natural History of Insects. 12mo. *London*, 1763. 26
The Natural History of Waters, Earths, Stones, Fossils, and Minerals.
12mo. *London*, 1763. 26

**Brougham** (Henry *Lord*). Dissertations on subjects of Science connected with Natural Theology...Paley's Work. 2 vols. 8vo. *London*,
1839. 12

**Broun** (*Capt.* Thomas). Manual of the New Zealand Coleoptera. 8vo.
*Wellington*, 1880. 95

**Brown** (Robert). The Miscellaneous Botanical Works of R. B., with
Atlas. 3 vols. (Ray Soc. Publ.) 8vo. & Fol. *London*, 1866–68.
17 & 72

**Browne** (J. Crichton). *See* West Riding Lunatic Asylum Reports.
Vol. 1–. 8vo. *London*, 1871–. 104

**Bruguières** (J. G.). Histoire naturelle des Vers (Encyclopédie méthodique). 2 tomes. 4to. *Paris*, 1789–92. 8

**Brunton** (T. Lauder). On Digitalis. 8vo. *London*, 1868. 92
*See* Klein (E.). Handbook for the Physiological Laboratory. 8vo. *London*, 1873. 115
Pharmacology and Therapeutics. 8vo. *London*, 1880. 92
The Bible and Science. 8vo. *London*, 1881. 26
\*The disorders of Digestion. 8vo. *London*, 1886. 104

**Buch** (L. von). Travels through Norway and Lapland...1806-8. Transl. by John Black. With Notes...by R. Jameson. 4to. *London*, 1813. **Ne**

**Bucke** (Richard Maurice). Man's moral nature. An Essay. 8vo. *London*, 1879. 12

**Buckler** (William). The larvæ of the British Butterflies and Moths. By (the late) W. B. Ed. by H. T. Stainton. 2 vols. (Ray Soc. Publ.) 8vo. *London*, 1886-87. 17a

**Buckley** (Arabella B.). A short history of Natural Science. 8vo. *London*, 1876. 26

**Buckton** (George Bowdler). Monograph of the British Aphides. 4 vols. (Ray Soc. Publ.) 8vo. *London*, 1876-83. 17a

**Büchner** (Ludwig). Aus Natur und Wissenschaft. 8vo. *Leipzig*, 1862. 28
Die Darwin'sche Theorie von der Entstehung und Umwandlung der Lebe-Welt. 4te Aufl. 8vo. *Leipzig*, 1876. 9
Sechs Vorlesungen über die Darwin'sche Theorie von der Verwandlung der Arten, &c. 8vo. *Leipzig*, 1868. 9
do. 2te Aufl. 8vo. *Leipzig*, 1868. 9
Conférences sur la Théorie Darwinienne. Traduit...par A. Jacquot. 8vo. *Leipzig*, 1869. 39
Die Macht der Vererbung. (Darwinistische Schriften, Nr. 12.) 8vo. *Leipzig*, 1882. 41
Die Stellung des Menschen in der Natur, &c. 2te Lief. Wer sind wir? 3te Lief. Wohin gehen wir? 8vo. *Leipzig*, 1870. 22
Man in the past, present and future. From the German of Dr L. B. By W. S. Dallas. 8vo. *London*, 1872. 124
Liebe und Liebes-Leben in der Thierwelt. 8vo. *Berlin*, 1879. 26
Mind in Animals. Transl. by Annie Besant. 8vo. *London*, 1880. 118

**Bütschli** (O.). Studien über die ersten Entwicklungsvorgänge der Eizelle ...der Infusorien. 4to. *Frankfurt a. M.*, 1876. **Nc**

**Buller** (Walter Lawry). A History of the Birds of New Zealand. 4to. *London*, 1873. 74

**Bulletin** de l'académie impériale des sciences de St Pétersbourg. Tomes XII. 4; XIII. 3, 5; XIV. 1; XV. 1–3; XVI. 5; XVII. 1; XVIII. 4; XIX. 1–4; XX. 3; XXI. 2, 3, 5; XXII.; XXIII. 2; XXV. 4, 5; XXVI. 2, 3; XXVII. 2, 4. 4to. 1867–81. 18

**Bulletin** de la soc. des sci. naturelles de Neuchâtel. Tome VI. 1, 3; VII. 2; VIII. 1–3; IX. 1; X. 3; XI. 1–3; XII. 1, 2. 8vo. 1863–81. 38

**Bulletins** de l'académie r. des sciences...de belgique. 2me Sér., Tomes 30–50, 1870–80. 8vo. *Bruxelles.* **79a**

**Burbidge** (F. W.). Cultivated Plants. 8vo. *Edinburgh,* 1877. **59**

**Burchell** (Wm J.). Travels in the Interior of Southern Africa. 2 vols. 8vo. *London,* 1822–24. **8**

**Burdon-Sanderson** (J.). *See* Klein (E.). Handbook for the Physiological Laboratory. 8vo. *London,* 1873. **115**

**Burgess** (Thomas H.). The Physiology or Mechanism of Blushing. 8vo. *London,* 1839. **39**

**Burmeister** (G.). *See* Anales del Museo Público de Buenos Aires. Tom. 1, 2. 4to. 1864–74. **72**

**Burmeister** (Hermann). Beiträge zur Naturgeschichte der Rankenfüsser (Cirripedia). 4to. *Berlin,* 1834. **Ne**
*See* Meyen (F. J. F.). Beiträge zur Zoologie. 4to. *Breslau,* 1834. **Ne**
The Organization of Trilobites. Ed. from the German by Th. Bell and Edw. Forbes. (Ray Soc. Publ.) Fol. *London,* 1846. **72**
Histoire de la Création...Trad. de l'allemand d'après la 8ème éd. par E. Maupas. Revue par le professeur Giebel. 8vo. *Paris,* 1870. **97**

**Busch** (Otto). Arthur Schopenhauer. Beitrag zu einer Dogmatik der Religionslosen. 8vo. *Heidelberg,* 1877. **113**
do. 2te Aufl. 8vo. *München,* 1878. **113**
Naturgeschichte der Kunst. 8vo. *Heidelberg,* 1877. **14**

**Busk** (George). Reports on Zoology for 1843, 1844. Transl. from the German by G. B., A. Tulk, and A. H. Haliday. (Ray Soc. Publ.) 8vo. *London,* 1847. **17**

**\*Butler** (Samuel). Evolution, old and new. (Op. 4.) 8vo. *London,* 1879. **40**
*See* Owen (J. P.). **24**

**Cabot** (J. Elliot). *See* Agassiz (L.). Lake Superior. 8vo. *Boston,* 1850. **15**

**Cabot** (Louis). The immature state of the Odonata. Part 1.—Subfamily Gomphina. Part 2.—Subfamily Æschnina. (Extr.) 4to. *Cambridge, Mass.,* 1871, 1881. **74**

**Calderwood** (Henry). Evolution and Man's place in Nature. 8vo. *London,* 1893. **11**

**\*Cambridge.** Studies from the Morphological Laboratory. Ed. by Adam Sedgwick. Vol. 2, Part 2 ; Vol. 3, Parts 1, 2. 8vo. *London,* 1886–88. **Ng**

**Camerano** (Lorenzo). La scelta sessuale. 8vo. *Torino,* 1880. **95**

**Cameron** (Peter). A monograph of the British Phytophagous Hymenoptera. Vols. 1, 2 (Ray Soc. Publ.). 8vo. *London,* 1882–85. **17a**

**Canada.** Geological Survey. Plans of various Lakes and Rivers between Lake Huron and the River Ottawa... (By A. Murray. Sir Wm. E. Logan, Director.) 4to. *Toronto*, 1857.      **74**

**Candolle** (Alphonse de). Géographie botanique raisonnée. 2 tomes. 8vo. *Paris*, 1855.      **60**
Histoire des sciences et des savants depuis deux siècles, &c. 8vo. *Genève*, 1873.      **13**
do. 2ᵉ éd. 8vo. *Genève-Bâle*, 1885.      **13**
La Phytographie. 8vo. *Paris*, 1880.      **57**
\*Origine des Plantes cultivées. 8vo. *Paris*, 1883.      **57**

**Candolle** (Alphonse et Casimir de). Monographiae Phanerogamarum. Prodromi nunc continuatio... 3 vols. 8vo. *Parisiis*, 1878–81.      **60**

**Candolle** (Augustin Pyramus de). Théorie élémentaire de la Botanique. 2de éd. 8vo. *Paris*, 1819.      **61**
Prodromus Systematis naturalis Regni Vegetabilis. Pars 1, 2. 2 vols. 8vo. *Parisiis*, 1824–25.      **62**
Mémoires et souvenirs de A. P. de C...Écrits par lui-même et publiés par son fils. 8vo. *Genève*, 1862.      **113**

**Canestrini** (Giovanni). Origine dell' Uomo. 2ª Ed. 8vo. *Milano*, 1870.      **10**
La Teoria dell' Evoluzione. 8vo. *Torino*, 1877.      **Nd**
La Teoria di Darwin criticamente esposta. 8vo. *Milano*, 1880.      **28**

**Carlier** (Antoine G.). Darwinism refuted by researches in Psychology. 8vo. *London* (1872).      **39**

**Carneri** (B.). Sittlichkeit und Darwinismus. 8vo. *Wien*, 1871.      **39**
Gefühl, Bewusstsein, Wille. 8vo. *Wien*, 1876.      **12**

**Carpenter** (William B.). Principles of Comparative Physiology. 4th ed. 8vo. *London*, 1854.      **125**
Researches on the Foraminifera. Part 1. (Extr.) 4to. *London* (1855).      **74**
Introduction to the study of the Foraminifera. By W. B. C., assisted by W. K. Parker and T. Rupert Jones. (Ray Soc. Publ.) Fol. *London*, 1862.      **72**
The Microscope and its revelations. 4th ed. 8vo. *London*, 1868.   **106**
Principles of Mental Physiology. 8vo. *London*, 1874.      **12**
*See* Jeffreys (G.). The ' Valorous' Expedition. Reports. (Extr.) 8vo. *London*, 1876.      **106**

**Carrière** (E. A.). Production et fixation des Variétés dans les végétaux. 8vo. *Paris* (1865).      **57**

**Carus** (J. Victor). Bibliotheca Zoologica. Verzeichniss der Schriften über Zoologie...1846-60. Bearb. von J. V. C., und W. Engelmann. 2 Bde. 8vo. *Leipzig*, 1861.      **96**
Geschichte der Zoologie bis auf Joh. Müller und Charl. Darwin. 8vo. *München*, 1872.      **112**
Handbuch der Zoologie. 1er Bd. ii. Hälfte. Wirbelthiere, Mollusken und Molluscoiden. 8vo. *Leipzig*, 1875.      **96**

**Carville** (H. C.). *See* Vulpian (A.).           **94**

**Caspari** (Otto). Die Urgeschichte der Menschheit. 2 Bde. 8vo. *Leipzig,* 1873.           **40**

**Catalogue**. Engelmann (Wilhelm). Bibliotheca Historico-Naturalis. 1ter Band. 8vo. *Leipzig,* 1846.           **10**
(Fossil Mammalia and Aves.) Descriptive and illustrated catalogue of the fossil organic remains of Mammalia and Aves contained in the museum of the Royal College of Surgeons of England. 4to. *London,* 1845.           **75**
(Geological Society.) Catalogue of the Books and Maps in the Library of the Geological Society of London. 8vo. *London,* 1846.    **107**
(Linnean Society.) Catalogue of the Natural History Library of the Linnean Society of London. Part 2. 8vo. *London,* 1867.    **35**
(Royal Society.) Catalogue of the Scientific Books in the Library of the Royal Society. 8vo. *London,* 1839.           **14**
of Scientific Papers (1800–83). Compiled and published by the Royal Society of London. 4to. Vols. 1–12.           **66**
*See* British Museum.
*See* Carus (J. V.) und W. Engelmann. Bibliotheca Zoologica. 1846–60. 2 Bde. 8vo. *Leipzig,* 1861.           **96**
*See* Ramsay (A. C.). A descriptive Catalogue of the Rock Specimens in the Museum of Practical Geology. 8vo. *London,* 1858.    **107**

**Caton** (John Dean). The Antelope and Deer of America. 8vo. *New York,* 1877.           **15**

**Cattaneo** (Giacomo). Darwinismo. Saggio sulla Evoluzione degli Organismi. 8vo. *Milano,* 1880.           **10**

**Cazelles** (E.). *See* Moleschott (Jac.). La circulation de la vie. Trad. 2 tomes. 8vo. *Paris,* 1866.           **11**

**Celsus** (A. C.). A. Corn. Celsi Medicinae Libri octo ex recensione Leonardi Targae...concinnavit E. Milligan. 8vo. *Edinb.* 1826.    **94**

**\*Chalon** (J.). Notes de Botanique expérimentale. 8vo. *Bruxelles,* 1897.           **60**

**Chambers** (Ch. Harcourt). *See* Gastaldi (B.). Lake Habitations. 8vo. *London,* 1865.           **124**

**Chambers** (Robert). Ancient Sea-margins. 8vo. *Edinburgh,* 1848.  **97**
Vestiges of the Natural History of Creation. 6th ed. 8vo. *London,* 1847.           **11**
\*do. 12th ed....Introd. by Alex. Ireland. 8vo. *London,* 1884.   **11**

**Chance** (Frank). *See* Virchow (R.). Cellular Pathology. 8vo. *London,* 1840.           **102**

**Chapman** (Henry C.). Evolution of Life. 8vo. *Philadelphia,* 1873.  **39**

**Chapman** (John). Neuralgia and kindred diseases of the Nervous System. 8vo. *London,* 1873.           **92**

**Chapuis** (F.). Le Pigeon voyageur belge. 8vo. *Verviers*, 1865.   127

**Charlesworth** (Edw.). *See* Magazine of Natural History. Vols. 1–4, N.S. 8vo. *London*, 1837–40.   **17a**

**Charpentier** (Jean de). Essai sur les Glaciers…du Bassin du Rhone. 8vo. *Lausanne*, 1841.   97

**Chaumont** (F. S. B. F. de). Lectures on State Medicine. 8vo. *London*, 1875.   93

**Child** (Gilbert W.). Essays on Physiological Subjects. 8vo. *Oxford*, 1868.   115
do. 2nd ed. 8vo. *London*, 1869. [2 copies.]   115

**Children** (John George). *See* Berzelius (J. J.). The use of the Blow-pipe. Transl. 8vo. *London*, 1822.   117
Memoir of J. G. Children. (Pr. pr.) 8vo. *Westminster*, 1853.   113

**Chun** (Carl). Fauna und Flora des Golfes von Neapel. 1. Monographie : Ctenophorae. 4to. *Leipzig*, 1880.   **Na**

*****Church** (Arthur H.). *See* Johnson (S. W.). How crops grow. 8vo. *London*, 1869.   59

**Clark** (Henry James). Mind in Nature. 8vo. *New York*, 1865.   12
Lucernariae and their allies. (Extr.) 4to. *Washington*, 1878.   **Na**

**Clarke** (Benjamin). On systematic Botany and Zoology, including a new arrangement of Phanerogamous Plants, &c. Oblong, *London*, 1870.   **Q. 2**

**Clarke** (J. W.). Cattle Problems explained. Thirty original Essays. 8vo. *Battle Creek, Mich.*, 1880.   108

**Claus** (Carl). Grundzüge der Zoologie. 2te vermehrte Aufl. 1–4 Lief. 8vo. *Marburg*, 1871–73.   96
Untersuchungen zur Erforschung der genealogischen Grundlage des Crustaceen-Systems. 4to. *Wien*, 1876.   72

**Cleland** (John). Evolution, Expression and Sensation, &c. 8vo. *Glasgow*, 1881.   12

*****Clements** (Frederic E.). *See* Pound (R.).   59

**Coan** (Titus). Adventures in Patagonia. 8vo. *New York*, 1880.   25

*****Coe** (Charles Clement). Nature *versus* Natural Selection. 8vo. *London*, 1895.   39

**Cognetti de Martiis** (S.). Le Forme primitive. 8vo. *Torino*, 1881.   39

**Cohn** (Ferdinand). *See* Henfrey (A.). Botanical and Physiological Memoirs. Transl. 8vo. *London*, 1853.   17
Die Pflanze. 8vo. *Breslau*, 1882.   61
*Blätter der Erinnerung zusammengestellt von seiner Gattin Pauline Cohn. Mit Beiträgen von Prof. F. Rosen. 8vo. *Breslau*, 1901.   123

18

**Colin** (G.).   Traité de Physiologie comparée des Animaux domestiques. 2 tomes.   8vo. *Paris*, 1854–56.   **125**

**Collett** (Robert).   Zoology.   Fishes.   (The Norwegian North-Atlantic Expedition, 1876–78.)   4to. *Christiania*, 1880.   **72**

**Collier** (James).   *See* Spencer (Herbert).   Descriptive Sociology.   English. Fol. *London*, 1873.   **Q. 2**

**Collingwood** (Cuthbert).   Rambles of a Naturalist on the shores and waters of the China Sea.   8vo. *London*, 1868.   **16**

**Collingwood** (J. F.).   *See* Waitz (Theodor).   Introduction to Anthropology.   8vo. *London*, 1863.   **124**

**Columbus** (Christopher).   Select Letters of C. C. .Transl. and ed. by R. H. Major.   (Hakluyt Soc. Publ.)   8vo. *London*, 1847.   **16**

**Comstock** (J. H.).   Report upon Cotton Insects.   8vo. *Washington*, 1879.   **102**
Report of the Entomologist of the U. S. Department of Agriculture for 1879.   Author's ed.   8vo. *Washington*, 1880.   **102**

**Congrès** international **d'Anthropologie** et d'Archéologie préhistoriques.   Compte rendu de la cinquième session à Bologne, 1871. 8vo. *Bologne*, 1873.   **124**

**Conta** (B.).   Théorie du Fatalisme.   8vo. *Bruxelles*, 1877.   **28**

**\*Conversations on Vegetable Physiology.**   2 vols.   12mo. *London*, 1829.   **62**

**Conybeare** (*Rev.* W. D.).   Outlines of the Geology of England and Wales.   By the Rev. W. D. C., and Wm. Phillips.   Part 1.   8vo. *London*, 1822.   **107**

**Cooke** (M. C.).   Mycographia, seu Icones Fungorum.   Vol. 1.   8vo. *London*, 1879.   **55**

**Cordillera de los Andes.**   Industria.   (Extr.)   8vo. [Philos. Tracts, i. 25.]   **11**

**Cotta** (Bernhard von).   Geognostische Wanderungen, ii.   8vo. *Dresden*, 1838.   **97**
Geology and History.   8vo. *London*, 1865.   **107**
Die Geologie der Gegenwart.   8vo. *Leipzig*, 1866.   **97**

**Cottage Gardener.**   Oct. 1855—Feb. 1856; Jan., Mar.—Nov. 1860; Apr. 1861.   4to. *London*.   **19**

**\*Coville** (Frederick V.).   Desert Botanical Laboratory of the Carnegie Institution.   By F. V. C., and D. T. Macdougal.   8vo. *Washington*, 1903.   **119**

**Cox** (Edward William).   What am I?   Vol. 1.   The Mechanism of Man. 8vo. *London*, 1873.   **12**

**\*Cramer** (Carl).  *See* Nägeli (C.).  Pflanzenphysiologische Untersuchungen.  1–4 Hefte.  4to. *Zürich*, 1855–58.  **Nc**

**\*Cramer** (Frank).  The Method of Darwin.  8vo. *Chicago*, 1896.  **13**

**Crawfurd** (John).  A Grammar and Dictionary of the Malay Language.
In 2 vols.  Vol. 1.  8vo. *London*, 1852.  **98**
A descriptive Dictionary of the Indian Islands, &c.  8vo. *London*, 1856.
**16**

**\*Crelle** (A. L.).  Rechentafeln welche alles Multipliciren und Dividiren
mit Zahlen unter Tausend ganz ersparen....  4te Stereotyp-Ausgabe.
Fol. *Berlin*, 1875.  **Na**

**Croll** (James).  Climate and Time in their geological relations.  8vo.
*London*, 1875.  **97**

**Crookes** (William).  Psychic Force and Modern Spiritualism.  8vo.
*London*, 1871.  **12**

**\*Crookes** (*Sir* William).  The Wheat Problem.  8vo. *London*, 1899.  **61**

**Cunningham** (Allan).  A few general remarks on the Vegetation of
certain coasts of Terra Australis, &c.  (Extr.)  [Philos. Tracts, i. 1.]
8vo.  **11**

**\*Cunningham** (D. D.).  i.  The causes of fluctuations in turgescence
in the motor organs of Leaves.  ii.  A new and parasitic species
of Choanephora.  Fol. *Calcutta*, 1895.  **Na**

**Cunningham** (Robert O.).  Notes on the Natural History of the Strait
of Magellan, &c.  8vo. *Edinburgh*, 1871.  **16**

**\*Cupples** (George).  Scotch Deer-hounds and their Masters.  By G. C.
With a biographical sketch of the author, by J. H. Stirling.  8vo.
*Edinburgh*, 1894.  **Nf**

**Currey** (Fr.).  *See* Schacht (H.).  **62**

**Curtis** (William).  The Botanical Magazine; or, Flower-Garden Displayed.  Vols. 1 and 2 (in 1 vol.).  8vo. *London*, 1793.  **59**

**Cuvier** (*le baron* G.).  Leçons d'anatomie comparée.  Rec. et publ....
par C. Duméril et G. L. Duvernoy.  Tomes 1–5.  8vo. *Paris*, An
VIII–XIV (1799–1805).  **116**
Essay on the Theory of the Earth.  With geological illustrations by
Professor Jameson.  5th ed.  Transl.  8vo. *Edinburgh*, 1827.  **97**
Le règne animal.  Nouv. éd., revue et augmentée.  Tomes 1–5.
(Tome 4, 5 = Crustacés, &c. par M. Latreille.)  8vo. *Paris*, 1829–30.
**108**

**Cyclopædia** (The) of Anatomy and Physiology.  Ed. by R. B. Todd.
5 vols. in 6.  8vo. *London*, 1859.  **Ng**

**Dallas** (W. S.).  *See* Büchner (L.).  Man in the past, present and future.
8vo. *London*, 1872.  **124**
A Natural History of the Animal Kingdom.  8vo. *London*, n. d.  **50**
*See* Siebold (C. T. E.).  **94**

**Dana** (James D.). On the Classification...of Crustacea. 4to. *Philadelphia*, 1853. 8
Manual of Geology. 8vo. *Philadelphia*, 1863. 50
Corals and Coral Islands. 8vo. *New York*, 1872. 50

**Dandolo** (*Count* V.). The art of rearing Silk-worms. Transl. from the work of Count Dandolo. 8vo. *London*, 1825. 95

**Danielssen** (D. C.) and Johan **Koren.** Zoology. Gephyrea. (The Norwegian North-Atlantic Expedition, 1876–78.) 4to. *Christiania*, 1881. 72

**Dareste** (Camille). Recherches sur la production artificielle des Monstruosités. 8vo. *Paris*, 1877. 23

**Darwin** (Charles). Extracts from Letters addressed to Professor Henslow by C. Darwin, Esq....printed for distribution among the Members of the Cambridge Philosophical Society. 8vo. *Cambridge*, 1835. [Philos. Tracts, ii. 4.] 11
A Letter...on the Moral State of Tahiti, New Zealand, &c. By Capt. R. FitzRoy and C. D. (Extr.) 8vo. At Sea, 28th June, 1836. [Philos. Tracts, ii. 3.] 11
The Zoology of the Voyage of H.M.S. Beagle...1832 to 1836. Edited and superintended by C. D. Parts 1–4, 4 vols. 4to. *London*, 1840–42. [2 copies.] 67
33 Photographs—Anthropological. **Na**
*See* Krause (Ernst). Erasmus Darwin. 41
*See* Romanes (G. J.). Mental Evolution in Animals...With a posthumous Essay on Instinct by C. D. 8vo. *London*, 1885. 47
*See* Weismann (A.). Studies in the Theory of Descent. 39

**Darwin** (Erasmus). *See* Krause (E.). 41

**Darwinian** (The) Theory of the Transmutation of Species examined by a Graduate of the University of Cambridge. 8vo. *London*, 1867. 39

**Darwinism.** The rise and influence of Darwinism. *See* Edinburgh Review (No. 402), Oct. 1902. 119

**Darwinistische Schriften.** Nr. 2–5 (2 copies), 6–9, 12. 8vo. *Leipzig*, 1878–82. 41

**Daubeny** (Charles). A description of active and extinct Volcanos. 8vo. *London*, 1826. 117

**Daubrée** (A.). Études...sur le métamorphisme et sur la formation des Roches cristallines. (Extr.) 4to. *Paris*, 1860. **Ne**

**Daubuisson** (J. F.). An account of the Basalts of Saxony. Translated, with Notes, by P. Neill. 8vo. *Edinburgh*, 1814. 107

*Davenport (C. B.). Statistical methods with special reference to Biological Variation. 1st ed. 8vo. *New York*, 1899. 106

Dawkins (W. Boyd). Cave Hunting. 8vo. *London*, 1874. 114

Dawson (James). Australian Aborigines. 4to. *Melbourne*, 1881. 75

Dawson (John W.). The Fossil Plants of the Devonian and Upper Silurian Formations of Canada. (Geolog. Survey of Canada.) 8vo. *Montreal*, 1871. 56

Defrance. Tableau des corps organisés Fossiles, &c. 8vo. *Paris*, 1824. 97

De la Beche (H. T.). A selection of the Geological Memoirs contained in the Annales des Mines, &c. Transl. with Notes. 8vo. *London*, 1824. 117
Researches in theoretical Geology. 8vo. *London*, 1834. 107

De la Cruz y Bahamonde (N.). *See* Molina (Juan I.). 24

Delage (Yves). Contribution à l'étude de l'appareil circulatoire des Crustacés édriophthalmes marins. 8vo. *Paris*, 1881. 75

Delamer (E. S.). Pigeons and Rabbits. 8vo. *London*, 1854. 127

Delgado (J. F. N.). Sobre a existencia do Terreno Siluriano no Baixo Alemtejo. (Com a trad. em francez.) 4to. *Lisboa*, 1876. 74

Delpino (Federico). Ulteriori osservazioni sulla Dicogamia nel regno vegetale. Parte 1ª; 2ª, Fasc. 1, 2. Estr. 8vo. *Milano*, 1868–74. 57

Demole (Isaac). *See* Heer (Oswald). 117

Denton (William). Is Darwin right? 8vo. *Wellesley, Mass.*, 1881. 23

*Derby (Edward Henry, 15th Earl of), K.G.* Speeches and Addresses on Political and Social Questions...1870–91. (Ed. by T. H. S.) 8vo. *London* (1893). 123
*Speeches and Addresses...Selected and edited by Sir T. H. Sanderson and E. S. Roscoe. 2 vols. 8vo. *London*, 1894. 123

Desmarest (A. G.). Mammalogie. 1ère, 2e Partie. 4to. *Paris*, 1820–22.

Devay (Francis). Du danger des mariages consanguins. 2e éd. 8vo. *Paris*, 1862. 11

Dictionary (*Dutch*). A new Pocket-Dictionary of the English and Dutch Languages. Stereotype ed. 8vo. *Leipzig*, 1878. 98
(*English*). Dictionary of the English Language. By Samuel Johnson. 2 vols. 4th ed. 8vo. *London*, 1770. 25
*(*French*). Le Dictionnaire royale, François-anglois, et Anglois-françois ...Par A. Boyer. Nouv. éd....par L. Du Mitand. 2 vols. (in 1 vol.). 4to. *London*, 1816. 67

(*German*). Rabenhorst's Pocket Dictionary of the German and English
Languages, in 2 Parts. By G. H. Noehden. 3rd ed. Rev. by
H. E. Lloyd. 24mo. *London*, 1829. 98
(*German*). A complete Dictionary of the English and German and
German and English Languages. In 2 vols. Vol. 1. Compiled...
by J. G. Flügel. 2nd ed. 8vo. *Leipzig*, 1838. 23
*(*Greek*). Græcum Lexicon Manuale primum a Benjamine Hederico
institutum...cura J. A. Ernesti. Ed. nova cui accedit magnus verb-
orum et exemplorum numerus ex schedis P. H. Larcheri. 4to.
*Londini*, 1816. 67
*(*Greek*). A new Greek and English Lexicon. By J. Donnegan.
3rd ed. 8vo. *London*, 1837. 14
*(*Latin*). Ainsworth's Latin Dictionary...New ed. by J. Carey. 4to.
*London*, 1816. 67
(*Spanish*). Neuman and Baretti's Dictionary of the Spanish and English
Languages. 5th ed. Rev. by M. Seone. In 2 vols. Vol. 1, Spanish
and English. 8vo. *London*, 1831. 14
*(*Welsh*). Welsh-English Dictionary...By G. ap Rhys. 8vo. *Carnarvon*
(1866). 98
(*Chemistry*). A Dictionary of Chemistry. 2nd ed. 8vo. *London*, 1823.
105
(*Chemistry*). A Dictionary of Chemistry. By Henry Watts. 5 vols.
and Supplement. 2nd ed. 8vo. *London*, 1871–72. 105
(*Gardener's*). *See* Miller (P.). 49

**Dictionnaire classique d'histoire naturelle.** Par Messieurs
Audouin, I. Boudin, &c. Tomes 1–17. 8vo. *Paris*, 1822–31. 41

**Dictionnaire des Sciences naturelles.** Par plusieurs Professeurs
du Jardin du Roi, &c. Tome 18ème. (Ga–Gju.) 8vo. *Paris*, 1820.
do. Planches. 1re, 2e Partie. 8vo. *Paris*, 1816–30. 41

**Dictionnaire**...des Termes usités dans les Sciences naturelles. Par
A.-J.-L. Jourdan. 2 tomes. 8vo. *Paris*, 1834. 41

**Dillwyn** (Lewis W.). A descriptive Catalogue of recent Shells. 2 vols.
8vo. *London*, 1817. 107

**Dippel** (Leopold). Das Mikroskop. 2ter Th. 8vo. *Braunschweig*, 1872.
106

**Dixie** (*Lady* Florence). Across Patagonia. 8vo. *London*, 1880. 15

**Dixon** (*Rev.* Edmund S.). Ornamental and Domestic Poultry. (Repr.)
8vo. *London*, 1848. 127
The Dovecot and the Aviary. 8vo. *London*, 1851. 127

**Dixon** (Frederic). The Geology and Fossils of the Tertiary and Cretaceous Formations of Sussex. 4to. *London*, 1850. 74

**Dobell** (Horace). Lectures on the germs and vestiges of Disease. 8vo.
*London*, 1861. 104

**Dobson** (George E.). *See* British Museum. Catalogue of the Chiroptera
in...the B. M. 8vo. *London*, 1878. 102

**Dodel** (Arnold). Die neuere Schöpfungsgeschichte nach dem gegenwärtigen Stande der Naturwissenschaften. 8vo. *Leipzig*, 1875. 57
Die Kraushaar-Alge, Ulothrix zonata. (Extr.) 8vo. *Leipzig*, 1876. 62

**Doherty** (Hugh). Philosophie organique. L'Homme et la Nature. 8vo. *Paris* (1881). 23

**Dohrn** (Anton). Untersuchungen über Bau und Entwicklung der Arthropoden. 1es Heft. 8vo. *Leipzig*, 1870. 106
Fauna und Flora des Golfes von Neapel. 3. Monographie. Pantopoda. 4to. *Leipzig*, 1881. **Na**

**Dollfus** (Gustave). Principes de Géologie transformiste. 8vo. *Paris*, 1874. 117

**Domestic Medicine.** A Hand-book...popularly arranged. By an eminent Physician. [By Δ.] 8vo. *London*, 1872. 104

**Donders** (F. C.). On the anomalies of Accommodation and Refraction of the Eye. Transl. by Wm. D. Moore. (New Sydenham Soc.) 8vo. *London*, 1864. 104

**Donn** (James). Hortus Cantabrigiensis. By the late J. D. Improved ...by Fr. Pursh. 10th ed., with additions, by J. Lindley. (Interleaved.) 8vo. *London*, 1823. 61

**\*Donnegan** (James). A new Greek and English Lexicon. 3rd ed. *See* Dictionary. 8vo. *London*, 1837. 14

**Douglas** (John Wm.). The British Hemiptera. Vol. 1. Hemiptera-Heteroptera. By J. W. D., and John Scott. (Ray Soc. Publ.) 8vo. *London*, 1865. 17

**Downing** (A. J.). The Fruits and Fruit Trees of America. 8vo. *London*, 1845. 49

**Drayson** (*Lieut.-Col.* A. W.). On the cause, date, and duration of the last Glacial Epoch of Geology, &c. 8vo. *London*, 1873. 107

**Dreher** (Eugen). Der Darwinismus und seine Stellung in der Philosophie. 8vo. *Berlin*, 1877. 39

**Drouët** (Henri). Mollusques marins des îles Açores. 4to. *Paris*, 1858. **Ne**

**\*Drude** (Oscar). Die Florenreiche der Erde. (Extr.) 4to. *Gotha*, 1884. **Nc**
\*Atlas der Pflanzenverbreitung. Fol. *Gotha*, 1887. **Q. 1**

**Drysdale** (John). The Protoplasmic Theory of Life. 8vo. *London*, 1874. 11

**Dub** (Julius). Kurze Darstellung der Lehre Darwin's über die Entstehung der Arten der Organismen. 8vo. *Stuttgart*, 1870. 39

**Du Bois-Reymond** (Emil). Gedächtnissrede auf Johannes Müller. (Extr.) 4to. *Berlin*, 1860. 8

**Duchenne** (G.-B.). Mécanisme de la Physionomie humaine. Avec un Atlas. 8vo. *Paris*, 1862. **75 & Q. 1**

**Dufrénoy** (P. A.). Mémoires pour servir à une description géologique de la France. Par MM. Dufrénoy et Élie de Beaumont. Tomes 3, 4. 8vo. *Paris*, 1836, 1838. **107**

**Du Fuchsia.** Par M. F. P\* \* \* \*. 8vo. *Paris*, 1844. **62**

**Dumont** (Léon A.). Haeckel et la Théorie de l'Évolution en Allemagne. 8vo. *Paris*, 1873. [2 copies.] **9**

**Duncan** (Andrew), jun. The Edinburgh New Dispensatory. 11th ed. 8vo. *Edinburgh*, 1826. **102**

**Duncan** (J. Matthews). Fecundity, fertility, sterility and allied topics. 2nd ed. 8vo. *Edinburgh*, 1871. **94**

**Duncan** (J. S.). Analogies of Organized Beings. 8vo. *Oxford*, 1831. **108**

**Dupont** (M. E.). L'Homme pendant les âges de la pierre dans les environs de Dinant-sur-Meuse. 8vo. *Bruxelles*, 1871. **114**

**Du Prel** (Karl *Freiherr*). Der Kampf um's Dasein am Himmel. 8vo. *Berlin*, 1874. **10**
do. 2te Aufl. 8vo. *Berlin*, 1876. **10**
Die Planetenbewohner und die Nebularhypothese. (Darwinistische Schriften, Nr. 8.) 8vo. *Leipzig*, 1880. **41**
Psychologie der Lyrik. (Darwinistische Schriften, Nr. 4.) 8vo. *Leipzig*, 1880. **41**

**Durand** (J.-P.) (de Gros). Essais de Physiologie philosophique. 8vo. *Paris*, 1866. **115**
Les origines animales de l'Homme, &c. 8vo. *Paris*, 1871. **39**

**Duval.** Histoire du Poirier (Pyrus sylvestris). 8vo. *Paris*, 1849. **62**
Histoire du Pêcher et sa culture. (Extr.) 8vo. *Paris*, 1850. **49**
Histoire du Pommier et sa culture. (Extr.) 8vo. *Paris*, 1852. **49**

**\*Dyer** (*Rev.* T. F. Thiselton). English Folk-Lore. 2nd ed. 8vo. *London*, 1880. **114**

**\*Dyer** (W. T. Thiselton). *See* Johnson (S. W.). How crops grow. 8vo. *London*, 1869. **59**
*See* Trimen (H.). Flora of Middlesex. 8vo. *London*, 1869. **62**

**Dzierzon.** Was ist die italienische Biene? *See* Bienen-Zeitung. 12ter Jahrg., Nro. 6, 1856. 4to. *Nördlingen*. **Ne**

**Eaton** (John M.). A treatise on...Tame, Domesticated, and Fancy Pigeons. 8vo. *London*, 1852. **127**
do. 1858. **127**
A treatise on...the Almond Tumbler, 1851. (Reprinted in the foregoing books.) **127**

**\*Ebermayer** (Ernst). Die gesammte Lehre der Waldstreu mit Rücksicht auf die chemische Statik des Waldbaues. 8vo. *Berlin*, 1876. 57

**Ecker** (Alexander). Die Anatomie des Frosches. 8vo. *Braunschweig*, 1864–82. 115

**Edgeworth** (M. Pakenham). Pollen. 8vo. *London*, 1877. 61

**Edinburgh** New Philosophical Journal. Vols. 24–43. 1837–47. 8vo. 38

**\*Edinburgh Review.** No. 402. 8vo. *London*, Oct. 1902. 119

**Edwards** (Milne). Histoire naturelle des Crustacés. Tomes 1–3 accompagnée de Planches. (Nouv. Suites à Buffon.) 8vo. *Paris*, 1834–40. 112

**Ehrenberg.** (A volume of Reprints on Infusoria, &c.) 8vo. *Berlin*, 1844–47. 106

**Ehrenberg** (Christian G.). Mikrogeologische Studien über das kleinste Leben der Meeres-Tiefgründe aller Zonen, &c. (Extr.) 4to. *Berlin*, 1873. 74

**Eichwald** (Eduard von). Geognostisch-palaeontologische Bemerkungen über die Halbinsel Mangischlak, &c. 8vo. *St Petersburg*, 1871. 117

**Eimer** (Th.). Untersuchungen über das Variiren der Mauereidechse. 8vo. *Berlin*, 1881. 97

**Elliott** (Henry W.). The Seal-Islands of Alaska. 4to. *Washington*, 1881. 74

**\*Ellis** (Ethel E.). Memoir of William Ellis. 8vo. *London*, 1888. 24

**Emery** (Carlo). Fauna und Flora des Golfes von Neapel. 2. Monographie : Fierasfer. 4to. *Leipzig*, 1880. **Na**

**Encyklopædie der Naturwissenschaften.** Hrsg. von G. Jäger (and others). 1te Abth., 1–29 Lief; 2te Abth., 1, 2 Lief. 8vo. *Breslau*, 1879–82. 21

**Endlicher** (Steph.). Prodromus Florae Norfolkicae…1804 et 1805 a F. Bauer collectae et depictae. 8vo. *Vindobonae*, 1833. 60

**Engelmann** (Wilhelm). *See* Carus (J. V.). 96
*See* Catalogue. Bibliotheca Historico-Naturalis. 1ter Band. 8vo. *Leipzig*, 1846. 10

**\*Engler** (Adolf). Versuch einer Entwicklungsgeschichte der Pflanzenwelt. 1, 2 Th. 8vo. *Leipzig*, 1879–82. 61

**Enten-** (Die) Schwanen- und Gänsezucht. (Probably by W. Riedel.) 8vo. *Ulm*, 1828. 127

**Entomological Society** of London. Transactions. Vol. 2, Part 4; Vol. 3, Parts 2–4; Vols. 4, 5; N.S., 1–5; 3rd ser., 1–5; and years 1868–81. 8vo. *London*, 1840–81. 36

**Ercolani** (G. B.). Sull' unità del tipo anatomico della Placenta nei Mammiferi e nell' Umana Specie, &c. 4to. *Bologna*, 1877. **74**
Nuove ricerche sulla Placenta nei Pesci cartilaginosi e nei Mammiferi, &c. 4to. *Bologna*, 1880. **74**

**\*Erichsen** (John E.). The Science and Art of Surgery. 5th ed. 2 vols. 8vo. *London*, 1869. **92**

**Erichson** (W.). *See* Meyen (F. J. F.). Beiträge zur Zoologie. 4to. *Breslau*, 1834. **Ne**

**Erichson** (W. F.). The Progress of Zoology in 1842. Transl. by W. B. Macdonald. (Ray Soc. Publ.—Reports...1841, 1842.) 8vo. *London*, 1845. **17**
Reports on Zoology for 1843, 1844. Transl. (Ray Soc. Publ.) 8vo. *London*, 1847. **17**

**Errera** (Léo). Sur la structure et les modes de Fécondation des Fleurs. Par L. E. et G. Gevaert. 1ère Partie. 8vo. *Gand*, 1878. **60**

**Eschricht** (D. F.). *See* Flower (W. H.). Recent Memoirs on the Cetacea. (Ray Soc. Publ.) Fol. *London*, 1866. **72**

**Eschwege** (W. L. von). Beiträge zur Gebirgskunde Brasiliens. 8vo. *Berlin*, 1832. **117**

**Espinas** (Alfred). Des sociétés animales. Étude de Psychologie comparée. 8vo. *Paris*, 1877. **47**

**\*Estlake** (Allan). The Oneida Community. 8vo. *London*, 1900. **28**

**Ewart** (J. Cossar). *See* Romanes (G. J.). Observations on...Echinodermata. 4to. *London*, 1881. **74**

**Eyton** (T. C.). Osteologia Avium. With Plates. 2 vols. 4to. *Wellington, Salop*, 1867. **75**

**Fabre** (J.-H.). Souvenirs entomologiques. 8vo. *Paris*, 1879. **102**

**Faivre** (Ernest). La variabilité des Espèces et ses limites. 8vo. *Paris*, 1868. **10**

**Falconer** (Hugh). Report on the Teak Forests of the Tenasserim Provinces. 8vo. *Calcutta*, 1852. **60**
Palæontological Memoirs and Notes of the late H. F.... Compiled and ed. by Charles Murchison. 2 vols. 8vo. *London*, 1868. **126**

**Falsan** (Albert). *See* Saporta (*le marquis* G. de). Recherches sur les végétaux fossiles de Meximieux. 4to. *Lyon*, 1876. **Q. 1**

**Farrar** (*Rev.* Fr. W.). Chapters on Language. 8vo. *London*, 1865. **14**

**Farrier and Naturalist.** Ed. by a Member of the Zoological Soc. of London. Vols. 1–3. 8vo. *London*, 1828–30. **108**

**Fayrer** (J.). The Royal Tiger of Bengal. 8vo. *London*, 1875. **118**

**\*Fenwick** (Samuel). The Student's Guide to Medical Diagnosis. 2nd ed. 8vo. *London*, 1871. **104**

**Ferguson** (George). Ferguson's illustrated series of rare and prize Poultry. 8vo. *London*, 1854. **127**

**Ferguson** (Robert). *See* Gooch on...Diseases peculiar to Women. 8vo. *London*, 1859. **94**

**Ferrière** (Émile). Le Darwinisme. (Bibl. utile.) 16mo. *Paris*, n. d. **23**
Le Darwinisme. 8vo. *Paris*, 1872. **10**

**Ferris** (B. G.). Origin of Species. A new Theory. 8vo. *Ithaca, N. Y.*, 1871. **23**

**Fichte** (Immanuel H.). Die Seelenfortdauer und die Weltstellung des Menschen. 8vo. *Leipzig*, 1867. **114**

**Finkler** (Dittmar). Ueber den Einfluss der Strömungsgeschwindigkeit und Menge des Blutes auf die thierische Verbrennung. (Extr. *See* Pflüger, E., Archiv f. Physiologie, Bd. x.) 8vo. *Bonn*, 1875. **105**

**\*Finland in the Nineteenth Century.** By Finnish Authors. Illustrated by Finnish Artists. (Editor, L. Mechelin.) Fol. *Helsingfors*, 1894. **Q. 1**

**\*Finland.** Pro Finlandia 1899—Les adresses internationales à S. M. l'Empereur-Grand-Duc Nicolas II. Fol. *London.* **Q. 1**

**Fischer** (P.). *See* Gaudry (A.). Animaux fossiles du mont Léberon. 4to. *Paris*, 1873. **Nb**

**Fiske** (John). Outlines of Cosmic Philosophy. 2 vols. 8vo. *London*, 1874. **41**
Darwinism and other Essays. 8vo. *London*, 1879. **10**

**Fitton** (William H.). Geological Notice on the new country passed over by Captain Back during his late Expedition [1833–34]. 8vo. [Philos. Tracts, i. 17.] **11**
Notes on the progress of Geology in England. (Extr.) 8vo. *London*, 1833. **97**

**Fitzgerald** (R. D.). Australian Orchids. Vol. 1 (7 Parts). Vol. 2, Parts 1, 3, 4, 5. Fol. *Sydney, N.S.W.*, 1877–. **Q. 1**

**FitzRoy** (*Admiral* Robert). Extracts from the Diary of an attempt to ascend the River Santa Cruz, in Patagonia...1834. (Extr.) 8vo. *London*, 1837. [Philos. Tracts, ii. 2.] **11**
A Letter...on the Moral State of Tahiti, New Zealand, &c. By Capt. R. FitzRoy, and Charles Darwin. (Extr.) 8vo. At Sea, 28th June, 1836. [Philos. Tracts, ii. 3.] **11**

**Fleming** (John). The Philosophy of Zoology. 2 vols. 8vo. *Edinburgh*, 1822. **112**
A History of British Animals. 8vo. *Edinburgh*, 1828. **118**

**Fleurieu** (C. P. Claret). Voyage autour du monde pendant les années 1790, 1791 et 1792, par Étienne Marchand. Précédé d'une introduction, &c. par C. P. C. F. Tomes 1–7. 8vo. *Paris*, An VI–VIII. 25

**Flint** (Charles L.). *See* Harris (T. W.). 102

**Flourens** (P.). De l'instinct et de l'intelligence des Animaux. 2de éd. 8vo. *Paris*, 1845. 11
De la Longévité humaine, &c. 8vo. *Paris*, 1854. 11
Examen du Livre de M. Darwin sur l'Origine des Espèces. 8vo. *Paris*, 1864. 10
*See* Blumenbach (J. F.). The Anthropological Treatises of J. F. B. (Memoir by Flourens.) Transl. 8vo. *London*, 1865. 114

**Flower** (*Sir* William H.). Recent Memoirs on the Cetacea by Professors Eschricht, Reinhardt and Lilljeborg. Ed. by W. H. F. (Ray Soc. Publ.) Fol. *London*, 1866. 72
An introduction to the Osteology of the Mammalia. 8vo. *London*, 1870. 108
Catalogue of the Specimens illustrating the Osteology and Dentition of Vertebrated Animals...in the Museum of the R. College of Surgeons of England. Part 1. Man: *Homo sapiens*, Linn. 8vo. *London*, 1879. 104

**Flügel** (J. G.). *See* Dictionary (German). 23

**Focke** (Wilhelm O.). Die Pflanzen-Mischlinge. 8vo. *Berlin*, 1881. 60

**Fol** (Hermann). Recherches sur la Fécondation...chez divers Animaux. (Extr.) 4to. *Genève-Bâle-Lyon*, 1879. 74

**Follen** (Charles). The life of C. F. By E. L. Follen. 8vo. *Boston U. S.*, 1844. 113

**Forbes** (Edward). On the Asteriadæ found fossil in British Strata. (Extr.) 8vo. *London* (1829). 117
A Monograph of the British Naked-eyed Medusæ. (Ray Soc. Publ.) Fol. *London*, 1848. 72

**Forbes** (W. A.). *See* Garrod (A. H.). The collected Scientific Papers of...A. H. G. 8vo. *London*, 1881. Nf

**Forel** (Auguste). Les Fourmis de la Suisse. 4to. *Zurich*, 1874. 75

**Forster** (John. Reinold). Observations made during a Voyage round the World, on Physical Geography, &c. 4to. *London*, 1778. 75

**Forster** (Thomas). A synoptical Catalogue of British Birds. 8vo. *London*, 1817. 127

**Fossils.** 22 loose Plates. Fol. Nb

**Foster** (*Sir* Michael). A course of elementary practical Physiology. Assisted by J. N. Langley. 8vo. *London*, 1872. [2 copies.]   115
*See* Klein (E.). Handbook for the Physiological Laboratory. 8vo. *London*, 1873.   115
The elements of Embryology. By M. F., and Francis M. Balfour. 8vo. *London*, 1874.   106
\**See* Balfour (F. M.). The Works of F. M. Balfour. 4 vols. Memorial Edition. 8vo. *London*, 1885.   Nd
\*A Text-Book of Physiology. 5th ed. Parts 1, 2. 2 vols. 8vo. *London*, 1888–89.   115

**Fournier** (Eugène). De la Fécondation dans les Phanérogames. 8vo. *Paris*, 1863.   62

**Fox** (Robert Were). Observations on Mineral Veins. (Extr.) 8vo. *Falmouth*, 1837. [Philos. Tracts, ii. 13.]   11

**Francisque-Michel.** Du passé et de l'avenir des Haras. 8vo. *Paris*, 1860.   108

**Freke** (H.). On the origin of species by means of Organic Affinity. 8vo. *London*, 1862.   39

**Frémont** (*Brev. Capt.* J. C.). Report of the exploring Expedition to the Rocky Mountains in the Year 1842, &c. 8vo. *Washington*, 1845.   25

**Frewer** (Ellen E.). *See* Holub (E.).   9

**Frey** (Heinrich). The Histology and Histochemistry of Man. By H. F. Transl. from the 4th German ed. by A. E. J. Barker. 8vo. *London*, 1874.   115

**Fritz** (Hermann). Die Beziehungen der Sonnenflecken zu den magnetischen und meteorologischen Erscheinungen der Erde. (Extr.) 4to. *Haarlem*, 1878.   Ne

**Frohschammer** (J.). Das Christenthum und die moderne Naturwissenschaft. 8vo. *Wien*, 1868.   28

**Fubini** (S.). *See* Moleschott (Jac.).   105

**Gallesio** (Georges). Traité du Citrus. 8vo. *Paris*, 1811.   61

**Galton** (Francis). The Narrative of an Explorer in Tropical South Africa. 8vo. *London*, 1853.   25
The Art of Travel. 8vo. *London*, 1855.   25
Hereditary Genius. 8vo. *London*, 1869.   13
English Men of Science. 8vo. *London*, 1874.   23
Inquiries into Human Faculty and its Development. 8vo. *London*, 1883.   22
\*Natural Inheritance. 8vo. *London*, 1889.   13

**Gamgee** (Arthur). *See* Hermann (D. L.). Elements of Human Physiology. Transl. 8vo. *London*, 1875.   115

**\*Garrod** (Alfred B.). The essentials of Materia Medica and Therapeutics. 3rd ed. 8vo. *London,* 1869. **104**

**Garrod** (Alfred Henry). The collected Scientific Papers of the late A. H. Garrod. Ed. by W. A. Forbes. (In Memoriam.) 8vo. *London,* 1881. **Nf**

**Gärtner** (Carl Friedrich v.). Versuche und Beobachtungen über die Befruchtungsorgane der vollkommeneren Gewächse. 8vo. *Stuttgart,* 1844. **57**
Versuche und Beobachtungen über die Bastarderzeugung im Pflanzenreich. 8vo. *Stuttgart,* 1849. **57**

**Gastaldi** (Bartolomeo). Lake Habitations...of Northern and Central Italy. Transl. and ed. by C. Harcourt Chambers. 8vo. *London,* 1865. **124**

**Gaudin** (Charles Th.). *See* Heer (O.). Recherches sur le climat et la végétation du pays tertiaire. 4to. *Winterthur,* 1861. **Q. 1**

**Gaudry** (Albert). Animaux fossiles et Géologie de l'Attique. 4to. *Paris,* 1862 [sheet].
Animaux fossiles et Géologie de l'Attique d'après les recherches faites en 1855–56 et en 1860. Avec un Atlas. Fol. *Paris,* 1862–67. **Q. 1**
Animaux fossiles du mont Léberon (Vaucluse)...Les Vertébrés par A. G....Les Invertébrés par P. Fischer et R. Tournouër. 4to. *Paris,* 1873. **Nb**
Les enchaînements du monde animal dans les temps géologiques. Mammifères tertiaires. 8vo. *Paris,* 1878. **Nf**

**Gaussoin** (Eugene). The Island of Navassa. Illustrated from Sketches. Obl. *Baltimore,* 1866. **Q. 2**

**Gay** (Claude). Aperçu sur les recherches d'histoire naturelle faites dans l'Amérique du Sud, &c....1830 et 1831. (Extr.) 8vo. *Paris,* 1833. [Philos. Tracts, ii. 10.] **11**

**\*Gee** (Samuel). Auscultation and Percussion. 8vo. *London,* 1870. **104**

**Gegenbaur** (Carl). Untersuchungen zur vergleichenden Anatomie der Wirbelthiere. 1tes Heft. 4to. *Leipzig,* 1864. **Ne**
Grundzüge der vergleichenden Anatomie. 2te Aufl. 8vo. *Leipzig,* 1870. **104**
Untersuchungen zur vergleichenden Anatomie der Wirbelthiere. 3tes Heft. Das Kopfskelet der Selachier. 4to. *Leipzig,* 1872. **Ne**
Manuel d'Anatomie comparée. Trad. en français sous la direction de Carl Vogt. 8vo. *Paris,* 1874. **55**
Elements of Comparative Anatomy. Transl. by F. Jeffrey Bell. The Transl. revised...by E. Ray Lankester. 8vo. *London,* 1878. **92**

**Geiger** (L.). Zur Entwickelungsgeschichte der Menschheit. 8vo. *Stuttgart,* 1871. **11**

**Geikie** (James). The great Ice Age. 2nd ed. 8vo. *London,* 1877. **97**
Prehistoric Europe. A Geological Sketch. 8vo. *London,* 1881. **107**

**Geographical Society.** See Royal.

**Geological Society.** Proceedings. Vol. 4, Parts 1, 2. Nos. 92–98.
8vo. *London,* 1842–44. **45**
Quarterly Journal. Vols. 1–37 and 38, Part 1, 1845–82. 8vo.
*London.* **45**
Transactions. 2nd Ser., Vol. 5, Part 1 ; Vol. 7, Part 2. 4to. *Lòndon,*
1837, 1845. **Nc**
*See* Catalogue of the Books and Maps. 8vo. *London,* 1846. **107**

**Geology.** The Future of Geology. *See* The Westminster Review, N.S.,
July, 1852. **119**

**Gérard** (R.). La fleur et le diagramme des Orchidées. 4to. *Paris,*
1879. **Nc**

**Gerland** (Georg). Ueber das Aussterben der Naturvölker. 8vo. *Leipzig,*
1868. **114**

**Gervais** (Paul). Histoire naturelle des Mammifères. 2 vols. 4to.
*Paris,* 1854–55. **Nd**

**\*Gibson** (R. J. H.). A Textbook of Elementary Biology. 8vo. *London,*
1889. **106**

**\*Gilbert** (J. H.). *See* Lawes (J. B.). Agricultural...experiments. 4to.
*London,* 1880. **Nc**

**Gillies** (John). An account of the Eruptions of the Volcano of Peu-
quenes, in the Andes of Chile. (Extr.) 8vo. *Edinburgh,* 1830.
[Philos. Tracts, ii. 11.] **11**

**Girton** (Daniel). The new and complete Pigeon-Fancier. New ed.
8vo. *London,* n. d. **117**

**\*Glazebrook** (R. T.). Light. (Cambridge Natural Science Manuals.)
8vo. *Cambridge,* 1894. **105**

**Glen** (Wm. Cunningham). The Poor Law Guardian. 2nd ed. 8vo.
*London,* 1857. **24**

**Gliddon** (Geo. R.). *See* Morton (S. G.). Types of Mankind. 8vo.
*Philadelphia,* 1854. **114**

**Gloger** (Constantin L.). Das Abändern der Vogel durch Einfluss des
Klima's. 8vo. *Breslau,* 1833. **127**

**Gmelin** (J. F.). *See* Linnæus (C.). Systema Naturæ. Ed. 13a. 8vo.
*Lugduni,* 1789–96. **61**

**Godman** (Fr. du Cane). Natural History of the Azores. 8vo. *London,*
1870. **25**

**Godron** (D. A.). De l'Espèce et des Races dans les êtres organisés.
2 tomes. 8vo. *Paris,* 1859. **40**

**Götz** (Theodor). Hunde-Galerie. Hrsg. von T. G. Obl. *Weimar,*
1853. **Na**

**Gonne** (Christian F.). Das Gleichgewicht in der Bewegung. 8vo. *Dresden*, 1882.     **28**

**Gooch** (Rob.). Gooch on some of the most important Diseases peculiar to Women, &c. Prefatory Essay by R. Ferguson. 8vo. *London*, 1859.     **94**

**Goodsir** (John and Harry D. S.). Anatomical and Pathological Observations. 8vo. *Edinburgh*, 1845.     **94**

**Gooseberry** (The) Growers' Register...for the year 1862. 12mo. *Macclesfield*.     **118**

**\*Gosse** (Edmund). The Life of Philip Henry Gosse, F.R.S. By his son E. G. 8vo. *London*, 1890.     **123**

**Gosse** (Philip H.). A Naturalist's Sojourn in Jamaica. By P. H. G. Assisted by Richard Hill. 8vo. *London*, 1851.     **41**
Letters from Alabama (U. S.). 8vo. *London*, 1859.     **26**
\*The Life of P. H. Gosse. *See* Gosse (E.).     **123**

**Gould** (Augustus A.). *See* Agassiz (L.). Principles of Zoölogy. Part 1. 8vo. *Boston*, 1848.     **106**

**Gould** (Benjamin A.). Investigations in the military and anthropological Statistics of American Soldiers. 8vo. *New York*, 1869.     **92**

**Gould** (John). Birds. (Zoology of...H.M.S. Beagle, Part 3.) 4to. *London*, 1841. [2 copies.]     **67**
An Introduction to the Birds of Australia. 8vo. *London*, 1848.     **127**
An Introduction to the Trochilidæ, or family of Humming-Birds. 8vo. *London*, 1861.     **127**
Handbook to the Birds of Australia. 2 vols. 8vo. *London*, 1865.     **117**
An Introduction to the Birds of Great Britain. 8vo. *London*, 1873.     **127**

**Gould** (*Rev*. William). An account of English Ants. 8vo. *London*, 1747.     **95**

**Graba** (Carl Julian). Tagebuch, geführt auf einer Reise nach Färö im Jahre 1828. 8vo. *Hamburg*, 1830.     **15**

**Grandeau** (Louis). *See* Gratiolet (P.).     **28**

**Grant** (*Rev*. Brewin). *See* Macaulay (J.). Vivisection. 8vo. *London*, 1881.     **28**

**Grant** (Robert E.). An essay on the study of the Animal Kingdom. 8vo. *London*, 1828. [Philos. Tracts, i. 2.]     **11**
Observations on the structure of some Silicious Sponges. (Extr.) 8vo. *Edinburgh* (1826). [Philos. Tracts, i. 3.]     **11**
Observations on the structure and functions of the Sponge. (Extr.) 8vo. *Edinburgh* (1825). [Philos. Tracts, i. 4.]     **11**
Observations on the structure and nature of Flustræ. (Extr.) 8vo. *Edinburgh*, 1827. [Philos. Tracts, i. 5.]     **11**

On the structure and characters of the Octopus ventricosus, Gr....from the Firth of Forth. (Extr.) 8vo. *Edinburgh*, 1827. [Philos. Tracts, i. 6.] 11

On...Ciliæ (*sic*) in the young of the Gasteropodous Mollusca, &c. (Extr.) 8vo. *Edinburgh*, 1827. [Philos. Tracts, i. 7.] 11

On...Lernæa elongata, Gr....from the Arctic Seas. (Extr.) 8vo. *Edinburgh* (1827). [Philos. Tracts, i. 8.] 11

Notice regarding the Ova of the Pontobdella muricata, Lam. (Extr.) 8vo. *Edinburgh* (1827). [Philos. Tracts, i. 9.] 11

Notice of a new Zoophyte (Cliona celata, Gr.) from the Frith of Forth. (Extr.) 8vo. *Edinburgh* (1826). [Philos. Tracts, i. 10.] 11

Observations on the Spontaneous Motions of the Ova of the Campanularia dichotoma, &c. (Extr.) 8vo. *Edinburgh*, 1826. [Philos. Tracts, i. 11.] 11

Outlines of Comparative Anatomy. Parts 1–4. [Incomplete.] 8vo. *London*, 1835–37. 94

**Gratiolet** (Pierre). De la Physionomie...suivi d'une notice sur sa vie, &c., par Louis Grandeau. 8vo. *Paris* (1865). 28

**Graves** (George). The Naturalist's Companion. 8vo. *Lond.*, 1824. 16

*****Gravis** (A.). Recherches...sur le Tradescantia virginica L. (Extr.) 4to. *Bruxelles*, 1898. **Nb**

**Gray** (Asa). Manual of the Botany of the Northern United States. 2nd ed. 8vo. *New York*, 1856. 62

First Lessons in Botany and Vegetable Physiology. 8vo. *New York*, 1857. 62

Botany for Young People. Part 2. How Plants behave. 4to. *New York*, 1872. 62

Natural Science and Religion. Two Lectures. 8vo. *New York*, 1880. 22

*Scientific Papers of Asa Gray. Selected by Charles Sprague Sargent. 2 vols. 1834–86. 8vo. *Boston*, 1889. 113

*Letters of Asa Gray. Ed. by Jane Loring Gray. 2 vols. 8vo. *Boston*, 1893. 113

**Gray** (George R.). Birds. 3 Parts. (The Zoology of the voyage of H.M.S. Erebus and Terror...1839–43.) 4to. *London*, 1844–45. 67

A fasciculus of the Birds of China. 4to. 1871. **Na**

*****Gray** (Henry). Anatomy descriptive and surgical. 5th ed., by T. Holmes. 8vo. *London*, 1869. **Nf**

**Gray** (John Edw.). Synopsis Reptilium. Part 1. Cataphracta. 8vo. *London*, 1831. 118

Mammalia. 3 Parts. (The Zoology of the voyage of H.M.S. Erebus and Terror...1839–43.) 4to. *London*, 1844–46. 67

Reptiles. 1 Part. (The Zoology of the Voyage of H.M.S. Erebus and Terror...1839–43.) 4to. *London*, 1845. 67

34

*See* British Museum. Catalogue...Mammalia. Parts 1–3. 12mo. *London*, 1843–52. 96

*See* British Museum. Catalogue of Marine Polyzoa. Parts 1, 2. 8vo. *London*, 1852–54. 96

Fauna of New Zealand (vol. 2, pp. 177–296). 8vo. 106

**Greene** (Joseph Reay). A manual of the sub-kingdom Cœlenterata. 8vo. *London*, 1861. 24

**Greenwell** (William). British Barrows. By W. G. Together with description of Figures of Skulls, &c. by George Rolleston. 8vo. *Oxford*, 1877. 114

**Greg** (Wm. Rathbone). The Creed of Christendom. 2nd ed. 8vo. *London*, 1863. 14

Enigmas of Life. 8vo. *London*, 1872. 12

**Grisebach** (A. H. R.). On Botanical Geography. Transl. by W. B. Macdonald and G. Busk. (Ray Soc. Publ.—Reports and Papers on Botany.) 8vo. *London*, 1846. 17

Die Vegetation der Erde nach ihrer klimatischen Anordnung. 2 Bde. 8vo. *Leipzig*, 1872. 58

**Grobben** (C.). Beiträge zur Kenntniss der männlichen Geschlechtsorgane der Dekapoden, &c. 8vo. *Wien*, 1878. (Extr.) 106

**Grote** (Arthur). *See* Hewitson (Wm. C.). Descriptions of New Indian Lepidopterous Insects. 4to. *Calcutta*, 1879. 74

**Grove** (W. R.). The correlation of Physical Forces. 4th ed. 8vo. *London*, 1862. 105

**Günther** (Albert C. L. G.). The Reptiles of British India. (Ray Soc. Publ.) Fol. *London*, 1864. 72

Description of Ceratodus, a genus of Ganoid Fishes. (Extr.) 4to. *London*, 1871. Ne

The gigantic Land-Tortoises (living and extinct) in the Collection of the British Museum. 4to. *London*, 1877. Na

An introduction to the study of Fishes. 8vo. *Edinburgh*, 1880. 106

**Gulliver** (George). *See* Hewson (Wm.). The Works of W. H. 8vo. *London*, 1846. 104

**Guthrie** (Malcolm). On Mr Spencer's Formula of Evolution. 8vo. *London*, 1879. 40

**\*Guy** (William A.). Principles of Forensic Medicine. 2nd ed. 8vo. *London*, 1861. 104

**\*Gwynne-Vaughan** (D. T.). *See* Bower (F. O.). 24

**Haast** (Julius von). Geology of the Provinces of Canterbury and Westland, New Zealand. 8vo. *Christchurch*, 1879. 107

**Haberlandt** (G.). Die Schutzeinrichtungen in der Keimpflanze. 8vo. *Wien,* 1877. 57
Vergleichende Anatomie des assimilatorischen Gewebesystems der Pflanzen. (Extr.) 8vo. *Berlin,* 1881. 60

**\*Hackley** (Charles E.). *See* Niemeyer (Felix von). 94

**Haeckel** (Ernst). Die Radiolarien (Rhizopoda radiaria). Mit einem Atlas. 2 vols. Fol. *Berlin,* 1862. **Q. 1**
Generelle Morphologie der Organismen. 2 Bde. 8vo. *Berlin,* 1866. 39
Zur Entwickelungsgeschichte der Siphonophoren. (Extr.) 4to. *Utrecht,* 1869. 74
Die Kalkschwämme. Eine Monographie in 2 Bänden Text und einem Atlas. 8vo. *Berlin,* 1872. **Nd**
Natürliche Schöpfungsgeschichte. 8vo. *Berlin,* 1868. 2te Aufl. 8vo. *Berlin,* 1870. 3te Aufl. 8vo. *Berlin,* 1872. 4te Aufl. 8vo. *Berlin,* 1873. 5te Aufl. 8vo. *Berlin,* 1874. 7te Aufl. 8vo. *Berlin,* 1879. 40
Anthropogenie oder Entwickelungsgeschichte des Menschen. 8vo. *Leipzig,* 1874. 3te Aufl. 8vo. *Leipzig,* 1877. 124
The History of Creation.... The Transl. revised by E. Ray Lankester. 2 vols. 8vo. *London,* 1876. 23
Arabische Korallen. 4to. *Berlin,* 1876. 72
Studien zur Gastraea-Theorie. 8vo. *Jena,* 1877. 106
Freie Wissenschaft und freie Lehre. 8vo. *Stuttgart,* 1878. 13
Das Protistenreich (Darwinistische Schriften, Nr. 1). 8vo. *Leipzig,* 1878. 41
Les preuves du Transformisme. Réponse à Virchow. Trad. par Jules Soury. 8vo. *Paris,* 1879. 11
Freedom in Science and Teaching. From the German of E. H. With a Prefatory Note by T. H. Huxley. 8vo. *London,* 1879. [2 copies.] 13
Das System der Medusen. 1te Theil. Mit einem Atlas. (4 Parts. Extr.) 4to. *Jena,* 1879-80. **Q. 1**
Gesammelte populäre Vorträge aus dem Gebiete der Entwickelungslehre. 2tes Heft. 8vo. *Bonn,* 1879. 40
The Evolution of Man. From the German of Ernst H. 2 vols. 8vo. *London,* 1879. 40

**Hagen** (Hermann A.). Monograph of the North American Astacidæ. (Extr.) 4to. *Cambridge, Mass.,* 1870. 74
On some insect deformities. (Extr.) 4to. *Cambridge, Mass.,* 1876. 74

**Hahn** (Otto). Die Urzelle. 8vo. *Tübingen,* 1879. 117
Die Meteorite (Chondrite) und ihre Organismen. 4to. *Tübingen,* 1880. 8
Die Philosophie des Bewussten. 8vo. *Tübingen,* 1887. 28

**\*Hales** (*Rev.* Steph.). Vegetable Staticks. 8vo. *London,* 1727. 61

**Hall** (John C.). *See* Pickering (Ch.). The Races of Man. New ed. 8vo. *London,* 1850. 124

3—2

**\*Hall** (Sidney). *See* Atlas. A new general Atlas...New ed., accompanied by an Alphabetical Index. Fol. *London,* n. d. Index. 8vo. *London,* 1831. **Q. 1 & Nd**

**Hallez** (Paul). Contributions à l'histoire naturelle des Turbellariés. 4to. *Lille,* 1879. **Ne**

**Hancock** (Albany). *See* Alder (J.). A Monograph of the British Nudibranchiate Mollusca. (Ray Soc. Publ.) Fol. *London,* 1845–55. **72**
On the Organization of the Brachiopoda. (Extr.) 4to. *London.* (Read 1857.) **Ne**

**Hansen** (Adolph). Vergleichende Untersuchungen über Adventivbildungen bei den Pflanzen. (Extr.) 4to. *Frankfurt a. M.,* 1881. **Nc**

**Harris** (George). The theory of the Arts. In 2 vols. Vol. 1. 8vo. *London,* 1869. **14**

**Harris** (Thaddeus W.). A treatise on some of the Insects of New England, which are injurious to Vegetation. 8vo. *Cambridge,* 1842. **102**
A treatise on some of the Insects injurious to Vegetation.... New ed. Ed. by C. L. Flint. 8vo. *Boston,* 1862. **102**
Entomological Correspondence of T. W. H. Ed. by Samuel H. Scudder. 8vo. *Boston,* 1869. **102**

**\*Hartig** (Robert). Lehrbuch der Baumkrankheiten. 8vo. *Berlin,* 1882. **60**

**Hartmann** (Eduard von). Wahrheit und Irrthum im Darwinismus. 8vo. *Berlin,* 1875. **39**

**Hartung** (Georg). Die geologischen Verhältnisse der Inseln Lanzarote und Fuertaventura. (Extr.) 4to. (*Zürich,* 1857). **74**

**Harvard College**, Cambridge, Mass. Bulletin of the Museum of Comparative Zoölogy. Vol. 2, No. 3 ; Vol. 3, Nos. 11—16 ; Vol. 5, Nos. 2—16 ; Vol. 6, Nos. 1—10, 12 ; Vol. 8, Nos. 1, 2, 4—14 ; Vol. 9, No. 6. 8vo. *Cambridge,* 1876–82. **105**
Illustrated Catalogue of the Museum of Comparative Zoölogy. [Incomplete. *See* Authors of Articles.] **74**

**Harvey** (William H.). Nereis Australis, or Algæ of the Southern Ocean. Part 2. 8vo. *London,* 1849. **Nc**
The Seaside Book. New ed. 8vo. *London,* 1849. **26**

**Hasse** (C.). Das natürliche System der Elasmobranchier. 4to. *Jena,* 1879. **Q. 1**

**Haughton** (*Rev.* Samuel). Six Lectures on Physical Geography. 8vo. *Dublin,* 1880. **89**

**Hawkins** (B. Waterhouse). A comparative view of the Human and Animal Frame. Fol. *London,* 1860. **Q. 1**

**Hawkins** (*Sir* Richard), *Knt.* The observations of Sir R. H., Knt., in his voyage into the South Sea...1593. Reprinted from the ed. of 1622. Ed. by C. R. Drinkwater Bethune. (Hakluyt Soc. Publ.) 8vo. *London*, 1847. **89**

**Head** (*Sir* F. B.), *Bart.* Rough notes taken during some rapid journeys across the Pampas and among the Andes. 8vo. *London*, 1826. **25**

**Heckel** (Édouard). Du mouvement végétal. 8vo. *Paris*, 1875. **57**

**\*Hedericus** (Benj.). *See* Dictionary (Greek). Græcum Lexicon Manuale ...Ed. nova. 4to. *Londini*, 1816. **67**

**Heer** (Oswald). Recherches sur le climat et la végétation du pays tertiaire. Par O. H., Traduction de Charles Th. Gaudin. 4to. *Winterthur*, 1861. **Q.1**
Untersuchungen über das Klima und die Vegetationsverhältnisse des Tertiärlandes. (Extr.) Fol. *Winterthur*, 1860. **Q.1**
Contributions to the Fossil Flora of North Greenland. (Extr.) 4tô. *London*. (Read 1869.) **Nd**
Die Miocene Flora und Fauna Spitzbergens. (Extr.) 4to. *Stockholm*, 1870. **Na**
Le Monde primitif de la Suisse. Trad. par Isaac Demole. 8vo. *Genève*, 1872. **117**
Flora fossilis Helvetiæ. Die vorweltliche Flora der Schweiz. 1–3 Lief. 4to. *Zürich*, 1876–77. **Nb**
Flora fossilis arctica. Die fossile Flora der Polarlände. Bde 3–5, 6, i. 4to. *Zürich*, 1875–80. **Na**
Pflanzenversteinerungen. (Extr.) 8vo. *Zürich*. **Na**
Ueber Ginkgo. (Extr.) 8vo. 1876? **Na**

**Heilprin** (Angelo). The geological evidences of Evolution. 8vo. *Philadelphia*, 1888. **10**

**Heliu.** La Loi unique et suprême. 1ère Partie. Genèse terrestre. 8vo. *Paris*, 1878. **22**

**Helmholtz** (H.). Popular Lectures on Scientific Subjects. Transl. by E. Atkinson. 8vo. *London*, 1873. **10**

**\*Hément** (Félix). *See* Mosso (A.). La Peur. 8vo. *Paris*, 1886. **22**

**Henfrey** (Arthur). Outlines of structural and physiological Botany. 8vo. *London*, 1847. **62**
*See* Mohl (Hugo von). Principles of the Anatomy...of the Vegetable Cell. Transl. 8vo. *London*, 1852. **122**
Botanical and Physiological Memoirs. (By A. Braun, G. Meneghini, and F. Cohn.) Translations. Ed. by A. Henfrey. (Ray Soc. Publ.) 8vo. *London*, 1853. **17**

**Henle** (J.). Handbuch der Muskellehre des Menschen. 8vo. *Braunschweig*, 1858. **94**

**Henry** (William). The elements of Experimental Chemistry. 9th ed. Vol. 2. 8vo. *London*, 1823. **105**

38

**Hensen** (V.). Physiologie der Zeugung. 8vo. *Leipzig,* 1881. 125

**Henslow** (*Rev.* John S.). 113
A Dictionary of Botanical Terms. New ed. 8vo. *London,* n. d. 62
*See* Darwin (Charles). Extracts from Letters addressed to Prof. H.
8vo. *Cambridge,* 1835. 11
The principles of descriptive and physiological Botany. New ed.
(Lardner's Cab. Cyclop.) 8vo. *London,* 1837. 128
Syllabus of Lectures on Botany, with an Appendix. 8vo. *Cambridge,*
1853. 62
*See* Jenyns (*Rev.* L.). 113

**Herbert** (*Hon.* and *Rev.* Wm.). Amaryllidaceæ. 8vo. *London,* 1837. 55

**Hermann** (H. C.). The Italian Alp-Bee. 8vo. *London,* 1860. 95

**Hermann** (L.). Elements of Human Physiology. Transl. from the 5th
ed. by A. Gamgee. 8vo. *London,* 1875. 115
Handbuch der Physiologie. 6ter Bd., ii Theil. *See* Hensen (V.),
Physiologie der Zeugung. 8vo. *Leipzig,* 1881. 125

**Herschel** (Caroline). Memoir and Correspondence of C. H. By
Mrs John Herschel. 8vo. *London,* 1876. 123

**Herschel** (*Sir* J. F. W.), *Knt.* A preliminary discourse on the study of
Natural Philosophy. (Lardner's Cab. Cyclop.) 8vo. *London,* 1831.
128
A treatise on Astronomy. (Lardner's Cab. Cyclop.) 8vo. *London,* 1833.
128
Physical Geography from the Encyclop. Britannica. 8vo. *Edinburgh,*
1861. 14

**Hervey-Saint-Denys** (*Le Baron* Léon). Recherches sur l'Agriculture
...des Chinois, &c. 8vo. *Paris,* 1850. 49

**Hewitson** (William C.). Descriptions of New Indian Lepidopterous
Insects...from the collection of the late Mr W. S. Atkinson. Rhopalocera, by W. C. H. Heterocera, by Fr. Moore. With an introd.
notice by Arthur Grote. 4to. *Calcutta,* 1879. 74

**Hewson** (William). The works of W. H. Ed. by G. Gulliver. (New
Sydenham Soc.) 8vo. *London,* 1846. 104

**Heyworth** (Lawrence). Glimpses at the Origin, Mission, and Destiny
of Man, &c. 8vo. *London,* 1866. 28

**Hibberd** (Shirley). The Fern Garden. 4th ed. 8vo. *London,* 1872. 62

**Higginson** (Th. Wentworth). Out-door Papers. 8vo. *Boston,* 1871. 26

**Hildebrand** (Friedrich). Die Verbreitung der Coniferen. (Extr.) 8vo.
*Bonn,* 1861. 60
Die Geschlechter-Vertheilung bei den Pflanzen, &c. 8vo. *Leipzig,* 1867.
57
Die Verbreitungsmittel der Pflanzen. 8vo. *Leipzig,* 1873. 57

**Hill** (Richard). *See* Gosse (Philip H.).      **41**

**Hinds** (Richard B.). The Regions of Vegetation. (Extr.) 8vo. *London*, 1843.      **59**

**Hippiatrist** (The). (Continuation of ' The Farrier and Naturalist.') Vol. 3. 8vo. *London*, 1830.      **108**

**Hitchcock** (Edward). Final Report on the Geology of Massachusetts. Vol. 2. 4to. *Northampton*, 1841.      **8**

**Hochstetter** (Ferdinand von). Reise der Österreichischen Fregatte Novara...1857–59, 1er Bd., 1te Abth., Geologie von Neu-Seeland; 2te Abth., Paläontologie von Neu-Seeland. 2 Bde. 4to. *Wien*, 1864–66.      **67**

**Hodge** (Charles), D.D. What is Darwinism? 8vo. *London*, 1874.      **10**

**Hodgson** (Shadworth H.). The Theory of Practice. An Ethical Enquiry. In two Books. 2 vols. 8vo. *London*, 1870.      **12**

**Hoek** (P. P. C.). Embryologie von Balanus. (Extr.) 8vo. *Leiden*, 1876.      **106**
Report on the Pycnogonida. (The Zoology of the Voyage of H.M.S. Challenger...1873–76, vol. 3, Part 10.) 4to. *London*, 1881.      **67**

**Hölder** (H. v.). Zusammenstellung der im Württemberg vorkommenden Schädelformen. 4to. *Stuttgart*, 1846.      **Na**

**Hoernes** (R.) und **Auinger** (M.). Die Gasteropoden der Meeres-ablagerungen der Ersten und Zweiten Miocänen Mediterran-Stufe. 1. Conus. (Extr.) 4to. *Wien*, 1879.      **72**
do. 3. Lief. Fol. *Wien*, 1882.      **Nb**

**Hofacker** (J. D.). Ueber die Eigenschaften welche sich bei Menschen und Thieren von den Eltern...vererben...mit Beiträgen von F. Notter. 8vo. *Tübingen*, 1828.      **108**

**Hoffmann** (H.). Zur Speciesfrage. (Extr.) 4to. *Haarlem*, 1875.      **Nc**

**Hoffmann** (L.). Thier-Psychologie. Bearb. von L. H. 8vo. *Stuttgart*, 1881.      **11**

**Hofmann** (A. W.). The life-work of Liebig. (The Faraday Lecture for 1875.) 8vo. *London*, 1876.      **123**

**Hofmeister** (Wilhelm). On the...Higher Cryptogamia, and...Coniferæ. Transl. by Fr. Currey. (Ray Soc. Publ.) 8vo. *London*, 1862.      **17**
Die Lehre von der Pflanzenzelle. 8vo. *Leipzig*, 1867.      **59**

*****Hogg** (Jabez). Elements of Experimental and Natural Philosophy. (2nd ed.) 8vo. *London*, 1861.      **105**

**Holland** (*Sir* Henry). Recollections of Past Life. 8vo. *London*, 1868.      **123**
Chapters on Mental Physiology. 8vo. *London*, 1852. 2nd ed. 8vo. *London*, 1858.      **22**

Medical Notes and Reflections. 8vo. *London,* 1839.     93
do. 3rd ed. 8vo. *London,* 1855.     93
Essays on Scientific and other subjects. 8vo. *London,* 1862.     16

**\*Holmes** (T.). *See* Gray (H.). Anatomy. 5th ed. 8vo. *London,* 1869.
    **Nf**

**Holub** (Emil). Seven years in South Africa. Transl. by Ellen E.
Frewer. 2nd ed. 2 vols. 8vo. *London,* 1881.     9
Beiträge zur Ornithologie Südafrikas von E. H. und Aug. von Pelzeln.
8vo. *Wien,* 1882.     117

**Homo** *versus* Darwin. 8vo. *London,* 1871. *See* Lyon (*Rev.* W. P.).     10

**\*Hooke** (Robert). Micrographia. 4to. *London,* 1667.     74

**Hooker** (*Sir* J. D.). i. On the Vegetation of the Carboniferous Period
as compared with that of the present day. ii. On some Peculiarities
in the Structure of *Stigmaria.* iii. Remarks on the Structure and
Affinities of some *Lepidostrobi.* (Mem. Geol. Survey of Gt. Britain,
Vol. 2, Part 2.)     56
The Botany of the Antarctic Voyage of H.M. discovery Ships *Erebus*
and *Terror*...1839–43. 4to. *London,* 1844–47.     **Nb**
Himalayan Journals. 2 vols. 8vo. *London,* 1854.     89
\*do. (Minerva Library.) 8vo. *London,* 1891.     89
Illustrations of Himalayan Plants chiefly selected from drawings made
for the late J. F. Cathcart, Esq. Plates by W. H. Fitch. Fol.
*London,* 1855.     **Q. 1**
Flora Indica. By J. D. H., and Th. Thomson. Vol. 1. 8vo. *London,*
1855.     59
On the Flora of Australia...being an introductory Essay to the Flora of
Tasmania. (Reprint.) 4to. *London,* 1859.     **Nb**
Introductory Essay to the Flora of New Zealand. (Reprint.) 4to. **Nb**
Botany. (Science Primers.) 8vo. *London,* 1876.     24
Journal of a Tour in Marocco and the Great Atlas. By J. D. H., and
John Ball...Including a sketch of the Geology of Marocco, by George
Maw. 8vo. *London,* 1878.     89

**Hooker** (William Dawson). Notes on Norway...1836. (Unpublished.)
2nd ed. 8vo. *Glasgow,* 1839.     15

**Hooker** (*Sir* W. J.). The British Flora. 2 vols. 4th ed. 8vo. *London,*
1838.     59
The British Flora. By Sir W. J. H., and G. A. W. Arnott. 7th ed.
8vo. *London,* 1855.     59

**Hope** (*Rev.* F. W.). The Coleopterist's Manual, containing the Lamelli-
corn Insects of Linneus and Fabricius. 8vo. *London,* 1837.     94

**Hopkins** (Evan). On the connexion of Geology with Terrestrial Mag-
netism. 8vo. *London,* 1844.     107

**Hopkins** (W.). An abstract of a Memoir on Physical Geology. (For
private circ.) 8vo. *Cambridge,* 1836. [Philos. Tracts, ii. 12.]     11

**Horner** (Leonard). On the occurrence of the Megalichthys in a Bed of Cannel Coal in the West of Fifeshire, &c. (Extr.) 8vo. *Edinburgh*, 1835. [Philos. Tracts, ii. 16.]  11
An account...of the alluvial land of Egypt. Part 2. (Extr.) 4to. *London*, 1858.  **Ne**

**Horticultural Society.** *See* Royal.

**Houghton** (*Rev.* W.). Gleanings from the natural history of the Ancients. 8vo. *London*, n. d.  114

**Hovelacque** (Abel). Notre Ancêtre. 2ème éd. 8vo. *Paris*, 1878.  124

**Howorth** (Henry H.). History of the Mongols from the 9th to the 19th Century. Part 1. 8vo. *London*, 1876.  **Nf**

**Hromada** (Adolf). Die vorsokratische Naturphilosophie der Griechen und die moderne Wissenschaft. (Separatabdruck.) 8vo. *Prag*, 1879.  23

**Huber** (François). Nouvelles observations sur les Abeilles. 2e éd. 2 tomes. 8vo. *Paris*, 1814.  95

**Huber** (Johannes). Die Lehre Darwin's kritisch betrachtet. 8vo. *München*, 1871.  10

**Huber** (P.). Recherches sur les mœurs des Fourmis indigènes. 8vo. *Paris*, 1810.  95

**Hubrecht** (A. A. W.). Studien zur Phylogenie des Nervensystems. 11. 4to. *Amsterdam*, 1882.  **Ne**

**Huc** (Évariste R.). Recollections of a Journey through Tartary, Thibet, and China...1844–46. A condensed Transl. by Mrs Percy Sinnett. 8vo. *London*, 1852.  10

**Hühner-** (Die) und Pfauenzucht. 8vo. *Ulm*, 1827.  127

**Humboldt** (Alexander von). Political Essay on the Kingdom of New Spain. Transl. from the original French, by John Black. 2 vols. 8vo. *New York*, 1811.  16
Personal Narrative of Travels to the Equinoctial Regions of the New Continent...1799–1804. By A. de H., and Aimé Bonpland...Transl. by Helen M. Williams. Vols. 1, 2, 3rd ed.; Vol. 3, 2nd ed.; Vols. 4, 5, 6, 7, 1st ed. In 6 vols. 8vo. *London*, 1819–.  15
Essai géognostique sur le Gisement des Roches dans les deux hémisphères. 2e éd. 8vo. *Paris*, 1826.  107
Fragmens de Géologie et de Climatologie asiatiques. 2 tomes. 8vo. *Paris*, 1831.  107
Cosmos. 2 vols. Translation. Ed. by Lt.-Col. Edw. Sabine. 8vo. *London*, 1846–48.  26
*See* Agassiz (L.). Address...Centennial Anniversary. 8vo. *Boston*, 1869.  113

**Hume** (David). Hume. By Prof. Huxley. 8vo. *London*, 1879.    113

**\*Humphreys** (George H.).    *See* Niemeyer (Felix von).    94

**Humphry** (*Sir* George Murray).    Observations on the Limbs of Vertebrate Animals. (Extr.)  4to. *Cambridge*, 1860.    75
\*Old Age and changes incidental to it.  8vo. *Cambridge*, 1885.    104

**Hunt** (James).  *See* Vogt (C.).  Lectures on Man.  8vo. *London*, 1864.
124

**Hunt** (Robert).   Researches on Light in its chemical relations.  2nd ed.
8vo. *London*, 1854.    105

**Hunter** (John).   The Natural History of the Human Teeth.  2nd ed.
4to. *London*, 1778.    **Ne**
*See* Abernethy (J.).  Physiological Lectures.  2nd ed.  8vo. *London*,
1822.    115
Memoranda on Vegetation.  4to. *London*, 1860.    **Ne**
*See* Blumenbach (J. F.).  The Anthropological Treatises of J. F. B...and
the Inaugural Dissertation of J. H. on the Varieties of Man.  Transl.
8vo. *London*, 1865.    114
Essays and Observations on Natural History, Anatomy, &c.... Arranged
and rev. with Notes, &c. by Richard Owen.  2 vols. 8vo. *London*,
1871.    13

**\*Huot** (J. J. N.).  *See* Atlas (de la Géographie) de Malte-Brun.  Fol.
*Paris*, 1837.    **Q. 1**

**Hutchinson** (*Lieut.-Col.* W. N.).  Dog breaking.  2nd ed.  8vo. *London*,
1850.    108

**Huth** (Alfred H.).   The Marriage of near Kin.  8vo. *London*, 1875.   40

**\*Hutton** (Frederick W.).   The Lesson of Evolution.  8vo. *London*, 1902.
59

**Hutton** (Thomas).   The Chronology of Creation.  8vo. *Calcutta*, 1850.
97

**Huxley** (Thomas H.).   The Oceanic Hydrozoa.  (Ray Soc. Publ.)  Fol.
*London*, 1859.    72
On our knowledge of the causes of the phenomena of Organic Nature.
Being six Lectures to Working Men, 1862.  8vo. *London*, 1863.
[2 copies.]    13
Evidence as to Man's place in Nature.  8vo. *London*, 1863.    40
Lectures on the Elements of Comparative Anatomy.  8vo. *London*, 1864.
115
An Introduction to the Classification of Animals.  8vo. *London*, 1869.
40
Lay Sermons, Addresses, and Reviews.  8vo. *London*, 1870.    23
A Manual of the Anatomy of Vertebrated Animals.  8vo. *London*, 1871.
106
Critiques and Addresses.  8vo. *London*, 1873.  [2 copies.]    13

Physiography. 8vo. *London*, 1877.     **14**
A Manual of the Anatomy of Invertebrated Animals. 8vo. *London*, 1877.
    **106**
*See* Haeckel (E.). Freedom in Science and Teaching. 8vo. *London*,
1879.     **13**
Hume. 8vo. *London*, 1879.     **113**
The Crayfish. 8vo. *London*, 1880.     **11**
Science and Culture and other Essays. 8vo. *London*, 1881.     **22**
*A course of practical instruction in Elementary Biology. By T. H. H.,
assisted by H. N. Martin. 8vo. *London*, 1875.     **106**
*A course of elementary instruction in Practical Biology. New ed.
8vo. *London*, 1881.     **106**
*Collected Essays. Vols. 1–9. 8vo. *London*, 1894.     **13**
American Addresses, &c. 8vo. *London*, 1877.     **13**
*See* The Quarterly Review, No. 385. Jan. 1901.     **113**

**Hyatt** (Alpheus). The genesis of the Tertiary Species of Planorbis
at Steinheim. (Extr.) 4to. *Boston*, 1880.     **74**

**Ichthyology.** (From the Encyclopædia Britannica, Vol. XII.) 4to.
*London*.     **Ne**

**Index** (The). A weekly Paper, devoted to Free Religion. Vol. I. Fol.
*Toledo, Ohio*, 1870.     **Na**

**Index Kewensis.** *See* Kew.     **Nc**

**Indian (The) Field.** Odd Numbers from Vols. 1–4. Fol. *Calcutta*,
1858–59.     **74**

**Ingersoll** (Ernest). The Oyster-Industry. [The Fisheries of the United
States.] 4to. *Washington*, 1881.     **74**

**Institut (L').** Journal général des Sociétés et Travaux scientifiques de
la France et de l'Étranger. 5e, 8e Année. 2 vols. 4to. *Paris*, 1837,
1840.     **Na**

**International Horticultural Exhibition,** and Botanical Congress,
held in London, from May 22nd to May 31st, 1866. Report of
Proceedings. 8vo.     **59**

**Irmisch** (Thilo). Beiträge zur Biologie und Morphologie der Orchideen.
4to. *Leipzig*, 1853.     **44**

*Jackson** (Benjamin Daydon). Vegetable Technology....Founded upon
the Collections of G. J. Symons. 4to. *London*, 1882.     **40**

**Jacquot** (Auguste). *See* Büchner (Louis).     **39**

**Jaeger** (Gustav). Die Darwin'sche Theorie und ihre Stellung zu Moral
und Religion. 8vo. *Stuttgart*, n. d.     **39**
Lehrbuch der allgemeinen Zoologie. 1, 2 Abth. 8vo. *Leipzig*, 1871–78.
    **96**
In Sachen Darwin's insbesondere contra Wigand. 8vo. *Stuttgart*, 1874.
    **23**

Seuchenfestigkeit und Constitutionskraft. 8vo. *Leipzig,* 1878.    **40**
Seuchenfestigkeit und Constitutionskraft. (Darwinistische Schriften, Nr. 2.) 8vo. *Leipzig,* 1878.    **41**
Zoologische Briefe. 8vo. *Wien,* 1876.    **96**

**James** (Constantin). Du Darwinisme ou l'Homme-singe. 8vo. *Paris,* 1877.    **11**

**Jameson** (Robert). A treatise on the external...characters of Minerals. 2nd ed. 8vo. *Edinburgh,* 1816.    **117**
Manual of Mineralogy. 8vo. *Edinburgh,* 1821.    **117**
*See* Buch (L. von).    **Ne**
*See* Cuvier (*Baron* G.). Essay on the Theory of the Earth. 5th ed. 8vo. *Edinburgh,* 1827.    **97**

**Jardine** (*Sir* William). *See* Agassiz (L.). Bibliographia Zoologiæ et Geologiæ (vol. 4). 8vo. *London,* 1854.    **17**

**Jarrold** (T.). Anthropologia: or, dissertations on the form and colour of Man. 4to. *London,* 1808.    **Ne**

**Jeffreys** (J. Gwyn). The 'Valorous' Expedition. Reports by J. G. Jeffreys and W. B. Carpenter. (Extr.) 8vo. *London,* 1876.    **106**

**Jenyns** (*Rev.* Leonard). A systematic Catalogue of British Vertebrate Animals. 8vo. *Cambridge,* 1835. [Philos. Tracts, i. 16.]    **11**
Fish. (Zoology of...H.M.S. Beagle, Part 4.) 4to. *London,* 1842. [2 copies.]    **67**
*See* White (*Rev.* Gilbert). The Natural History of Selborne. New ed. 8vo. *London,* 1843.    **10**
Observations in Natural History. 8vo. *London,* 1846.    **26**
Observations in Meteorology. 8vo. *London,* 1858.    **26**
Memoir of the Rev. John Stevens Henslow. 8vo. *London,* 1862.    **113**
*See also* Blomefield (Leonard) (late Jenyns).

**\*Jevons** (W. Stanley). Elementary Lessons in Logic. New ed. 8vo. *London,* 1881.    **24**

**Johnson** (Charles and C. P.). *See* Sowerby (J. E.). British Poisonous Plants. 2nd ed. 8vo. *London,* 1861.    **62**

**Johnson** (Samuel). *See* Dictionary.    **25**

**\*Johnson** (Samuel W.). How Crops grow. Revised...by A. H. Church and W. T. Thiselton Dyer. 8vo. *London,* 1869.    **59**

**Johnston** (George). The Botany of the Eastern Borders. 8vo. *London,* 1853.    **61**

**\*Johnston** (R.). Civil Service Guide. 13th ed. 8vo. *Lond.,* 1893. **24**

**Jones** (John Matthew). The Naturalist in Bermuda. By J. M. J.... assisted by Maj. J. W. Wedderburn and J. L. Hurdis. 8vo. *London,* 1859.    **25**

**Jordan** (Alexis). De l'origine des diverses variétés ou espèces d'arbres fruitiers, &c. *See* Mém. de l'acad. imp....de Lyon. Classe des sciences. (Nouv. sér.) Tome 2ème. 8vo. *Lyon,* 1852. **44**

**Joseph** (Domingo). *See* Molina (Juan I.). **24**

**Jourdan** (A. J. L.). *See* Dictionnaire. **41**

**Journal of Anatomy and Physiology.** Vol. 3, Part 2 (2nd ser., Vol. 2) to Vol. 8 (2nd ser., Vol. 7); Vols. 11–15; Vol. 16, Part 1. 8vo. *Cambridge,* 1869–. **Nh**

**Journal of Anthropology.** Vol. 1, Nos. 1–3. 8vo. *London,* 1870–71. **42**

**Journal of Horticulture.** Apr. 1861—Feb. 1866 [some Nos. missing]. 4to. *London.* **19**

**Journal of a Horticultural Tour** through some parts of Flanders... 1817. By a Deputation of the Caledonian Horticultural Society. 8vo. *Edinburgh,* 1823. **61**

**Journal of Travel and Natural History.** Ed. by Andrew Murray. Vol. 1. 8vo. *London,* 1868–69. **9**

**Jouvencel** (Paul de). Genèse selon la Science. La Vie. 8vo. *Paris,* 1859. **11**

**Juan** (George). A voyage to South America...undertaken...by G. Juan and A. de Ulloa, Captains of the Spanish Navy. Transl. by John Adams. 4th ed. Vol. 1. 8vo. *London,* 1806. **95**

**Judd** (John W.). Volcanoes, what they are and what they teach. 8vo. *London,* 1881. **11**

**Jukes** (J. Beete). The Student's Manual of Geology. 8vo. *Edinburgh,* 1857. **50**

**Kaspary** (Joachim). Natural Laws; or the Infallible Criterion. 8vo. *London,* 1876. New ed. 8vo. *Edinburgh,* 1862. **23**

**Kater** (*Capt.* Henry). A treatise on Mechanics. By Capt. H. K., and *Rev.* D. Lardner. (Lardner's Cab. Cyclop.) 8vo. *London,* 1830. **128**

**Keir** (James). Sketch of the life of James Keir, Esq., F.R.S. (Pr. pr.) 8vo. *London* (1859). **123**

**Keiserling.** *See* Keyserling.

**Kerner** (A.). Die Cultur der Alpenpflanzen. 8vo. *Innsbruck,* 1864. **62** Flowers and their unbidden guests. The transl. rev. and ed. by W. Ogle. 8vo. *London,* 1878. **59**

**\*Kew.** Index Kewensis. 4 Parts. 4to. *Oxonii,* 1893–95. **Nc** *\*See* British Museum and Kew. Committee on Botanical Work. Report, and Minutes of Evidence...dated 11th March, 1901. Fol. *London.* **Nc**

**Key** (Axel). *See* Retzius (G.). **Q. 3**

**Keyserling** (Alexander *Graf* von). Wissenschaftliche Beobachtungen auf einer Reise in das Petschora-Land im Jahre 1843. (Hrsg. von Alex. *Graf* Keyserling.) Pp. 1–288. 4to. *St Petersburg*, 1846. **Ne**

**Keyserling** (Alex. *Graf* von) and P. von **Krusenstern**. 1. Geognostisch-Geographische Uebersicht des Petschora Landes. 1846. 2. Karte der Flüsse Petschora, &c. 1846. **Ne**

**Kickx** (Jean Jacques). *See* Strasburger (Ed.). Sur la formation...des Cellules. 8vo. *Jena*, 1876. 60

**Kidd** (William). The Canary. Cheap ed. 8vo. *London*, n. d. 127

**King** (Phillip Parker), *Capt.*, R.N. Some Observations upon the Geography of the Southern Extremity of South America, &c....1826–30. (Extr.) 8vo. *London*, 1831. [Philos. Tracts, ii. 5.] 11

Description of the Cirrhipedia, Conchifera and Mollusca in a collection formed...1826–1830 in surveying the Southern Coasts of South America, &c. By P. P. K., and W. J. Broderip. (Zool. Jl., v., 1832–34.) 8vo. *London*. [Philos. Tracts, ii. 6.] 11

**Kirby** (*Rev.* William). Monographia Apum Angliæ, &c. 2 vols. 8vo. *Ipswich*, 1802. 95
An introduction to Entomology. By Rev. W. K., and Wm. Spence. Vol. 1, 3rd ed.; Vol. 2, 2nd ed.; Vols. 3 and 4, 1st ed. 8vo. *London*, 1818–26. 112
*See* Richardson (J.). Fauna Boreali-Americana. 3 Parts. 4to. *London*, 1829–36. 75

**Kirchhof** (F.). Das Ganze der Landwirthschaft. Hrsg. von F. K. 13tes Heft: Die Schweine- und Geflügelzucht. 8vo. *Leipzig*, 1835. 108

**\*Kirchner** (O.). Flora von Stuttgart und Umgebung. 8vo. *Stuttgart*, 1888. 62

**Kirkes** (Wm. Senhouse). *See* Baly (Wm.). Recent advances in the Physiology of Motion, &c. 8vo. *London*, 1848. 125

**Klein** (E.). Handbook for the Physiological Laboratory. By E. K., J. Burdon-Sanderson, M. Foster, and T. Lauder Brunton. Ed. by J. Burdon Sanderson. 2 vols. 8vo. *London*, 1873. 115
The Anatomy of the Lymphatic System. 1. The Serous Membranes. 2. The Lungs. 4to. *London*, 1873–75. 92

**Kobell** (Franz von). Grundzüge der Mineralogie. 8vo. *Nürnberg*, 1838. 117

**Kölliker** (A.). Anatomisch-systematische Beschreibung der Alcyonarien. 1te Abt. Die Pennatuliden. 1te Hälfte. 4to. *Frankfurt a. M.*, 1870. **Ne**

**Kölreuter** (Joseph G.). Vorläufige Nachricht von einigen das Geschlecht der Pflanzen betreffenden Versuchen. 8vo. *Leipzig*, 1761–66. 61

**Körner** (Friedrich). Thierseele und Menschengeist. 8vo. *Leipzig*, 1872.
12

**\*Kohlrausch** (F.). Leitfaden der praktischen Physik. 3te Aufl. 8vo.
*Leipzig*, 1877. 95

**Koninck** (L. de) et H. **Le Hon**. Recherches sur les Crinoïdes du
terrain carbonifère de la Belgique, &c. 4to. *Bruxelles*, 1854. **Ne**

**Koren** (Johan). *See* Danielssen (D. C.). 72

**Kosmos.** 1877–1886. 8vo. *Leipzig und Stuttgart.* 21

**Kowalevsky** (Waldemar). Monographie der Gattung Anthracotherium
Cuv., &c. 1er Th. 4to. *Cassel*, 1873. 74

**Krause** (Ernst). *See* Allen (Grant). Der Farbensinn. 8vo. *Leipzig*,
1880. 41
Erasmus Darwin und seine Stellung in der Geschichte der Descendenz-
Theorie. Von E. K. Mit seinem Lebens- und Charakterbilde von
Charles Darwin. (Darwinistische Schriften, Nr. 6.) 8vo. *Leipzig*,
1880. 41
\*Hermann Müller von Lippstadt. Ein Gedenkblatt. 8vo. *Lippstadt*,
1884. 113
*See* Sterne (Carus). 26

**Krusenstern** (Paul von). Geographische Ortsbestimmungen. *See* Keyserling (A. *Graf* von). Wissenschaftliche Beobachtungen...Pet-
schora-Land. 4to. *St Petersburg*, 1846. **Ne**

**Kühne** (H.). Die Bedeutung des Anpassungsgesetzes für die Therapie.
(Darwinistische Schriften, Nr. 3.) 8vo. *Leipzig*, 1878. 41

**Kühne** (W.). Untersuchungen über das Protoplasma und die Contracti-
lität. 8vo. *Leipzig*, 1864. 57

**Kuhl** (Joseph). Die Descendenzlehre und der neue Glaube. 8vo.
*München*, 1879. 39

**\*Kunkel** (A. J.). Handbuch der Toxikologie. 1te Hälfte. 8vo. *Jena*,
1899. **Nf**

**Kuntze** (Otto). Methodik der Speciesbeschreibung und Rubus. 4to.
*Leipzig*, 1879. **Nc**
Um die Erde. 8vo. *Leipzig*, 1881. 15

**Kurr** (Johann G.). Untersuchungen über die Bedeutung der Nektarien
in den Blumen. 8vo. *Stuttgart*, 1833. 62

**Kurtz** (F.). *See* Munk (H.). Die elektrischen und Bewegungs-Erscheinungen am Blatte der Dionaea muscipula. 8vo. *Leipzig*, 1876. 122

**Labillardière** (J. J. de). Relation du voyage à la recherche de la
Pérouse...1791–. 2 vols. 8vo. *Paris*. An VIII. 15

**Lacepède** (*Le Cte de*). Histoire naturelle des Cétacées. 2 tomes.
(In 1 vol.) 8vo. *Paris*, 1809. 118

**Lacordaire** (J. Th.). Mémoire sur les habitudes des Coléoptères de l'Amérique méridionale. (Extr.) 8vo. *Paris* (1830). [Philos. Tracts, ii. 17.] 11

**Laing** (Sidney H.). Darwinism refuted. 8vo. *London*, 1871. 11

**Laird** (James L.). *See* Wagner (Moritz). The Darwinian Theory, &c. Transl. 8vo. *London*, 1873. 39

**Lamarck** (J. B. P. A.). Histoire naturelle des Animaux sans vertèbres. tomes 1–7. 8vo. *Paris*, 1815–22. 50
do. tomes 1–11. 2ème éd. 8vo. *Paris*, 1835–45. 50
Philosophie zoologique. Nouv. éd. tome 1er. 8vo. *Paris*, 1830. 112
do. Nouv. éd., revue...par C. Martins. 2 tomes. 8vo. *Paris*, 1873. 112

**Lambert** (Charles). Le système du Monde moral. 8vo. *Paris*, 1862. 28
L'Immortalité selon le Christ. 8vo. *Paris*, 1865. 28

**Lambertye** (*Le Comte* Léonce de). Le Fraisier. 8vo. *Paris* (1863). 60

**Lamont** (James). Seasons with the Sea-Horses. 8vo. *Lond.*, 1861. 26

**Lamouroux** (J.). Exposition méthodique des genres de l'ordre des Polypiers. 4to. *Paris*, 1821. 67

**Lanciano** (Raffaele). L' Universo l' Astro e l' Individuo. 8vo. *Napoli*, 1872. 12

**Landois** (H.). *See* Altum (B.). 106

**Lanessan** (J. L. de). Du Protoplasma végétal. 4to. *Paris*, 1876. 44
Étude sur la doctrine de Darwin. La Lutte pour l'Existence, &c. 8vo. *Paris*, 1881. 11

**Langley** (J. N.). *See* Foster (M.). A course of elementary practical Physiology. 8vo. *London*, 1872. 115

**Lankester** (Edwin). *See* Ray (John). Memorials of J. R. 8vo. *London*, 1846. 17
*See* Macgillivray (Wm.). The Natural History of Dee Side and Braemar. (Pr. pr.) 8vo. *London*, 1855. 16

**Lankester** (E. Ray). On comparative longevity in man and the lower animals. 8vo. *London*, 1870. 9
Degeneration. 8vo. *London*, 1880. 13
*See* Gegenbaur (Carl). 92

**Lardner** (*Rev.* Dionysius). *See* Kater (*Capt.* H.) and *Rev.* D. L. A treatise on Mechanics. (The Cabinet Cyclop.) 8vo. *London*, 1830. 128
The Cabinet Cyclopædia. [A selection—Eleven Vols.] 8vo. *London*. 128

**\*Latham** (P. W.). The Harveian Oration...Oct. 18, 1888. 8vo. *Cambridge*, 1888.     **14**

**Latham** (R. G.). Man and his migrations. 8vo. *London*, 1851.   **114**

**Latreille** (P. A.). Histoire naturelle des Fourmis, &c. 8vo. *Paris*, An x—1802.     **95**
*See* Cuvier. Le Règne animal. Nouv. éd. (See tomes 4, 5.) 8vo. *Paris*, 1829–30.     **108**

**Laugel** (Aug.). Science et Philosophie. 8vo. *Paris*, 1863.     **11**
Les problèmes de la Nature. 8vo. *Paris*, 1864.     **11**

**Lavater** (Gaspard). L'art de connaître les Hommes par la Physionomie. 10 tomes. 8vo. *Paris*, 1820.     **38**

**\*Lawes** (J. B.). Agricultural, Botanical, and Chemical results of experiments on the mixed herbage of permanent meadow. Part 1. By J. B. L., and J. H. Gilbert. (Extr.) 4to. *London*, 1880.   **Nc**

**Lawrence** (John). The Horse in all his varieties and uses. 8vo. *London*, 1829.     **118**

**Lawrence** (W.). Lectures on Physiology, Zoology, and the Natural History of Man. 8vo. *London*, 1822.     **115**

**Lawson** (Peter) and Son, *Seedsmen*. Lists of Seeds, Plants, &c., &c. 4to. *Edinburgh*, 1851.     **62**

**Le Brun** (Charles). The Conference of Monsieur Le Brun...upon Expression. Transl. from the French. 8vo. *London*, 1701.   **128**

**Lecky** (W. E. H.). *See* Derby (Edw. H., 15th *Earl of*).     **123**

**Lecoq** (Henri). De la Fécondation...des Végétaux et de l'Hybridation. 8vo. *Paris*, 1845.     **59**
do. 2ème éd. 8vo. *Paris*, 1862.     **59**
Études sur la Géographie botanique de l'Europe. Tomes 1–9. 8vo. *Paris*, 1854–58.     **60**

**Le Couteur** (John). On...Wheat. 8vo. *Jersey*, n. d.     **49**
do. 2nd ed. 8vo. *London*, 1872.     **49**

**Lees** (William). Elements of Acoustics, Light, and Heat. 8vo. *London*, 1877.     **128**

**Le Hon** (H.). L'Homme fossile en Europe. 8vo. *Bruxelles*, 1867.   **124**
*See* Koninck (L. de).     **Ne**

**Leidy** (Joseph). The Ancient Fauna of Nebraska. (Extr.) 4to. *Washington*, 1853.     **Na**
Contributions to the Extinct Vertebrate Fauna of the Western Territories. [U. S. Geological Survey.] 4to. *Washington*, 1873.   **74**

**Leighton** (*Rev.* W. A.). The Lichen-Flora of Great Britain, Ireland and the Channel Islands. 2nd ed. 12mo. *Shrewsbury*, 1872.   **62**

**Lemoine** (Albert). De la Physionomie et de la Parole. 8vo. *Paris,* 1865. 28

**Lemoine** (Victor). Recherches sur les Oiseaux fossiles des terrains tertiaires inférieurs des environs de Reims. 8vo. *Reims,* 1878. 75

**Lepelletier de la Sarthe** (A.). Traité complet de Physiognomonie. 8vo. *Paris,* 1864. 28

**Leslie** (Alex.). *See* Nordenskiöld (A. E.). 15

**Lesquereux** (Leo). Report on the Fossil Plants of the auriferous gravel deposits of the Sierra Nevada. (Extr.) 4to. *Cambridge, Mass.,* 1878. 74

**Lesson** (René-Primevère). Manuel de Mammalogie. 12mo. *Paris,* 1827. 118
Manuel d'Ornithologie. 2 tomes. 12mo. *Paris,* 1828. 118

**Letourneau** (Ch.). Physiologie des Passions. 8vo. *Paris,* 1868. 12

**Levier** (Émile). *See* Schiff (M.). 125

*****Lewes** (George H.). The Physiology of Common Life. Vol. 2. 8vo. *Edinburgh,* 1860. 115
The History of Philosophy. 3rd ed. 2 vols. 8vo. *London,* 1867. 28
The Physical Basis of Mind. 8vo. *London,* 1877. 28

**Leybold** (Federico). Escursion a las Pampas Arjentinas...Febrero de 1871. 8vo. *Santiago,* 1873. 9

**Lhotsky** (*Dr* John). A journey from Sydney to the Australian Alps... 1834. 8vo. *Sydney,* 1835. [Philos. Tracts, i. 22.] (Incomplete.) 11

**Liebig** (Justus von). Organic Chemistry in its applications to Agriculture and Physiology. Ed... by Lyon Playfair. 8vo. *London,* 1840. 24
*See* Hofmann (A. W.). The Life-work of Liebig. 8vo. *London,* 1876. 123

**Lilljeborg** (W.). *See* Flower (W. H.). Recent Memoirs on the Cetacea. (Ray Soc. Publ.) Fol. *London,* 1866. 72

**Lindemuth** (H.). Vegetative Bastarderzeugung durch Impfung. (Extr.) 8vo. *Berlin,* 1878. 61

**Lindley** (John). *See* Donn (J.). Hortus Cantabrigiensis. 10th ed. (Interleaved.) 8vo. *London,* 1823. 61
A Natural System of Botany. 2nd ed. 8vo. *London,* 1836. 61
School Botany and Vegetable Physiology. New ed. 8vo. *London,* 1856. 61

*****Lindsay** (B.). An introduction to the study of Zoology. 8vo. *London,* 1895. 106

**Link** (H. F.).   Die Urwelt und das Alterthum, erläutert durch die Natur-
kunde.   1er Th.   8vo.   *Berlin,* 1821.                    23
The Progress of Physiological Botany in 1841.   Transl. by E. Lankester.
(Ray Soc. Publ.   Reports...1841, 1842.)   8vo.   *London,* 1845.    17
Report on Botany.   Transl. by J. Hudson.   (Ray Soc. Publ.)   8vo.
*London,* 1846.                                            17

**Linnæus** (Carl).   Philosophia Botanica.   Ed. 2a.   8vo.   *Viennæ,* 1783.
                                                            61
*Caroli Linnæi, sveci, Doctoris medicinæ, **Systema Naturæ.**   A
facsimile of the first edition (Leyden, 1735) published by the R.
Swedish Academy.   Folio, *Stockholm,* 1907.               **Q. 3**
Systema Naturæ.   Ed. 13a.   Cura J. F. Gmelin.   (Bd. in 10 vols.)
8vo.   *Lugduni,* 1789–96.                                 61
Systema Vegetabilium.   Ed. 15a quae ipsa est recognitionis a b. Io. A.
Murray institutae tertia, procurata a C. H. Persoon.   8vo.   *Gottingae,*
1797.                                                       61
*A general view of the writings of Linnæus, by Richard Pulteney.
2nd ed., by William G. Maton.   4to.   *London,* 1805.     **Nc**

***Linnéporträtt*** vid Uppsala universitets minnefest på tvåhundraårsdagen
af Carl von Linnés födelse: af Tycho Tullberg.   4to.   *Stockholm,* 1907.
                                                            **Na**

**Linnean Society.**   *See* Catalogue of the Natural History Library.
Part 2.   8vo.   *London,* 1867.                           35
Journal of the Proceedings.   Botany and Zoology.   Vols. 1–8.   8vo.
*London,* 1856–64.   Journal.   Botany.   Vols. 9–37.   8vo.   *London,*
1865–1906.   Journal.   Zoology.   Vols. 9–30.   8vo.   *London,* 1866–
1907.                                                       34–35, 43
Proceedings, 1869–1906.   8vo.   *London.*                 34
do.   List.   1875–1907.                                    34
*do.   Charter and Bye-laws.   Nov. 1904.   8vo.           34
Transactions.   Vol. 21, Parts 3, 4; 22–30.   1854–.   General Index to
vols. 1–30.   2nd Series—Botany.   Vols. 1–6; 7, Parts 1–5.   4to.
1875–.                                                      65
Transactions.   2nd Ser.   Zoology.   Vol. 1; 2, Parts 1–3, 13–17;
Vols. 3–6; 7, (Part 5 missing); 8, Parts 1–13; 9, Parts 1–11;
10, Parts 1–7.   4to.   1875–.                             65

***Linsbauer*** (Karl).   *See* Wiesner und seine Schule.   Festschrift.   8vo.
*Wien,* 1903.                                               35

***Linsbauer*** (Ludwig).   *See* Wiesner und seine Schule.   Festschrift.
8vo.   *Wien,* 1903.                                        35

**Lippert** (Julius).   Die Religionen der Europäischen Culturvölker...in
ihrem geschichtlichen Ursprunge.   8vo.   *Berlin,* 1881.  24

**Lisle** (Edward).   Observations in Husbandry.   2 vols.   2nd ed.   8vo.
*London,* 1757.                                             49

**Locard** (Arnould). Études sur les variations malacologiques...du bassin du Rhone. 2 tomes. 8vo. *Lyon*, 1881. **Nf**

**Logan** (*Sir* Wm. E.). *See* Canada. Geological Survey. **74**

**Loiseleur-Deslongchamps** (J. L. A.). Considérations sur les Céréales. (Partie historique.) (Extr.) 8vo. *Paris*, 1842. **59**

**Lombardini** (Luigi). Sui Cammelli. 4to. *Pisa*, 1879. **Ne**

**Loudon.** Arboretum Britannicum. Accidents, Diseases, and Insects. (Extr.) 8vo. [Philos. Tracts, i. 13.] **11**

**Lovén** (S.). Études sur les Échinoïdées. Texte avec Planches. (Extr.) 4to. *Stockholm*, 1875. **74**

**Low** (David). On the Domesticated Animals of the British Islands. 8vo. *London*, 1845. **118**

**Lowne** (Benjamin Thompson). The anatomy and physiology of the Blow-fly. 8vo. *London*, 1870. **102**
The Philosophy of Evolution. 8vo. *London*, 1873. **9**

**Lowthorp** (John). *See* Royal Society of London. The Philosophical Transactions and Collections to the end of the year 1700. Abridg'd. In three vols. 4to. *London*, 1705. **71a**

**Lubbock** (*Sir* John), *Bart.* An account of the two methods of Reproduction in *Daphnia*, &c. (Extr.) 4to. *London*, 1857. **74**
On the Ova and Pseudova of Insects. (Extr.) 4to. *London* (Read 1858). **Ne**
Pre-Historic Times, as illustrated by ancient remains, &c. 8vo. *London*, 1865. **114**
do. 2nd ed. 8vo. *London*, 1869. **114**
*See* Nilsson (Sven). The Primitive Inhabitants of Scandinavia. 3rd ed. 8vo. *London*, 1868. **124**
The Origin of Civilisation, &c. 8vo. *London*, 1870. **124**
Monograph of the Collembola and Thysanura. (Ray Soc. Publ.) 8vo. *London*, 1873. **17**
On the origin and metamorphoses of Insects. (Nature Series.) 8vo. *London*, 1874. **102**
Addresses, political and educational. 8vo. *London*, 1879. **40**
Scientific Lectures. 8vo. *London*, 1879. **40**
Chapters in Popular Natural History. 8vo. *London* (1882). **26**
Ants, Bees and Wasps. 4th ed. 8vo. *London*, 1882. **11**
*The Senses, Instincts, and Intelligence of Animals. 8vo. *London*, 1888. **11**

**Lucae** (Johann C. G.). Zur Statik und Mechanik der Quadrupeden (Felis und Lemur). (Herrn Geh. Sanitätsrath Dr Georg Varrentrapp ...fünfzigjährigen Doctor-Jubiläums.) 4to. *Frankfurt a. M.*, 1881. **72**

Der Fuchs-Affe und das Faulthier...in ihrem Knocken- und Muskel-skelet. (Herrn Geh. Dr Ludwig W. T. v. Bischoff...fünfzigjährigen Doctor-Jubiläums.) 4to. *Frankfurt a. M.*, 1882. **72**

**Lucas** (Prosper). Traité philosophique et physiologique de l'Hérédité naturelle. 2 tomes. 8vo. *Paris,* 1847–50.                          13

**Lütken** (Chr. Fred.). *See* Steenstrup (J. J. S.).                          Ne

**Lunze** (Gustav). Die Hundezucht im Lichte der Darwin'schen Theorie. 8vo. *Berlin,* 1877.                          40

**Lyell** (*Sir* Charles), *Bart.* Principles of Geology. 3 vols. 8vo. *London,* 1830–33.                          103
 do. 5th ed. 4 vols. 12mo. *London,* 1837.                          103
 do. 6th ed. 3 vols. 12mo. *London,* 1840.                          103
 do. 7th ed. 8vo. *London,* 1847.                          103
 do. 9th ed. 8vo. *London,* 1853. [2 copies.]                          103
 do. 10th ed. 2 vols. 8vo. *London,* 1867–68. [2 copies.]                          103
 do. 11th ed. 2 vols. 8vo. *London,* 1872.                          103
On the Cretaceous and Tertiary Strata of the Danish Islands of Seeland and Möen. *See* Trans. Geol. Soc. of London. 2nd Ser., Vol. 5, Part 1. 4to. *London* [Read 1835].                          Nc
Elements of Geology. 12mo. *London,* 1838.                          113
 do. 6th ed. 8vo. *London,* 1865.                          113
Travels in North America; &c. 2 vols. 8vo. *London,* 1845.                          25
A second visit to the United States of North America. 2 vols. 8vo. *London,* 1849.                          25
A Manual of Elementary Geology. 3rd ed. 8vo. *London,* 1851.                          113
 do. 4th ed. 8vo. *London,* 1852.                          113
 do. 5th ed. 8vo. *London,* 1855.                          113
 do. Supplement to the 5th ed. 1st and 2nd ed. 8vo. *London,* 1857.                          113
 do. 6th ed. [See " Elements of Geology."]                          113
The Geological Evidences of the Antiquity of Man. 8vo. *London,* 1863.                          113
 do. 4th ed. 8vo. *London,* 1873.                          113
The Student's Elements of Geology. 8vo. *London,* 1871.                          113
 do. 2nd ed. 8vo. *London,* 1874.                          113

**Lyell** (James C.). Fancy Pigeons. 8vo. *London,* 1881.                          117

**Lyman** (Theodore). Supplement to the Ophiuridæ and Astrophytidæ. (Extr.) 4to. *Cambridge, Mass.,* 1871.                          74
Ophiuridæ and Astrophytidæ (of the Hassler Expedition). (Extr.) 4to. *Cambridge, Mass.,* 1875.                          74

[**Lyon** (*Rev.* W. P.).] Homo *versus* Darwin. 8vo. *London,* 1871.                          10

**McAlpine** (D.). Zoological Atlas. Vertebrata. Obl. *Edinburgh,* 1881.                          Q. 1
The Botanical Atlas. Part 1. 4to. *Edinburgh,* 1882.                          72

**Macaulay** (James). Vivisection…Prize Essays. By J. M., Rev. B. Grant, and A. Wall. 8vo. *London,* 1881.                          28

**McClelland** (John). "Indian Cyprinidæ." [Asiatic Researches, 19th vol., Part 2.] 4to. *Calcutta,* 1839.                          74

**MacClintock** (*Sir* F. L.).   The voyage of the 'Fox' in the Arctic Seas.
A Narrative of the discovery of the fate of Sir John Franklin, &c.
8vo. *London*, 1859.   **41**

**MacCulloch** (John).   A Geological Classification of Rocks.   8vo.
*London*, 1821.   **97**

**\*Macdougal** (Daniel T.).   *See* Coville (Fr. V.).   **119**

**Macgillivray** (William).   A History of British Birds.   5 vols.   8vo.
*London*, 1837–52.   **117**
The Natural History of Dee Side and Braemar.   Ed. by E. Lankester.
(Pr. pr.) 8vo. *London*, 1855.   **16**

**Mackintosh** (*Sir* James).   The History of England.   Vol. 1.   (Lardner's
Cab. Cyclop.)   8vo. *London*, 1830.   **128**
Dissertation on the progress of Ethical Philosophy…With a Preface by
the Rev. Wm. Whewell.   2nd ed.   8vo. *Edinburgh*, 1837.   **12**

**McIntosh** (W. C.).   A Monograph of the British Annelids.   Part 1.
The Nemerteans.   Fol. *London*, 1873–74.   **72**

**Maclaren** (James).   A critical examination of some of the principal
arguments for and against Darwinism.   8vo. *London*, 1876.   **10**
Natural Theology in the 19th Century.   8vo. *London*, 1878.   **12**

**Maclaren** (J. J.).   Some chemical difficulties of Evolution.   8vo.
*London*, 1877.   **23**

**McLennan** (John F.).   Primitive Marriage.   8vo. *Edinburgh*, 1865.
**114**
Studies in Ancient History comprising a reprint of 'Primitive Marriage,'
&c.   8vo. *London*, 1876.   **114**

**McNab** (Wm. R.).   Botany.   Outlines of Classification of Plants.   8vo.
*London*, 1878.   [2 copies.]   **62**

**Macquart** (J.).   Facultés intérieures des Animaux invertébrés.   8vo.
*Lille*, 1850.   **106**

**Magazine of Natural History,** and Journal of Zoology, &c.   Conducted by J. C. Loudon.   Vols. 1–9.   8vo. *London*, 1829–36.   **43a**

**Magazine of Natural History.**   Conducted by Edward Charlesworth.   Vols. 1–4, N. S.   8vo. *London*, 1837–40.   **17a**

**Magazine of Zoology and Botany.**   Conducted by Sir W. Jardine,
*Bart.*, P. J. Selby and Dr Johnston.   Vols. 1, 2.   8vo. *Edinburgh*,
1837–38.   **Nh**

**Maillet** (Benoît de).   Telliamed: or discourses…on the diminution of
the Sea…Transl. from the French.   8vo. *London*, 1750.   **23**

**Major** (R. H.).   *See* Columbus (Chr.).   Select Letters.   8vo. *London*,
1847.   **16**

**Mallery** (Garrick). Introduction to the study of Sign Language among the North American Indians. 4to. *Washington*, 1880.          75

A collection of Gesture-signs and Signals of the North American Indians. (Distributed only to Collaborators.) 4to. *Washington*, 1880.          75

Sign Language among North American Indians compared with that among other peoples and deaf-mutes. (Extr.) 8vo. *Washington*, 1881.          75

**Malm** (A. W.). Göteborgs och Bohusläns Fauna, Ryggradsdjuren. 8vo. *Göteborg*, 1877.          **Nf**

**\*Malpighius** (M.). Anatome Plantarum. Cui subjungitur Appendix... de Ovo incubato observationes continens. Fol. *Londini*, 1675.          **Na**

**\*Malte-Brun.** *See* Atlas (de la Géographie). Fol. *Paris*, 1837.          **Q. 1**

**Malthus** (*Rev.* T. R.). An Essay on the principle of Population. 6th ed. 2 vols. 8vo. *London*, 1826. [2 copies.]          39

**Mantegazza** (Paolo). Rio de la Plata e Tenerife. 8vo. *Milano*, 1867.          25

Fisiologia del Piacere. 5a ed. 8vo. *Milano*, 1870.          22

Studii antropologici ed etnografici sulla Nuova Guinea. 8vo. *Firenze*, 1877.          114

Il terzo molare nelle razze umane. 8vo. (*Firenze*, 1878).          40

Fisonomia e Mimica. 8vo. *Milano*, 1881.          28

La physionomie et l'expression des sentiments. 8vo. *Paris*, 1885.          28

**Map.** Military Map of the United States. 1870. Scale 1 : 5,000,000.          **Q. 3**

Map of the States of Kansas and Texas and Indian Territory, &c. 1867. Scale 1 : 1,500,000.          **Q. 3**

Territory of the United States from the Mississippi River to the Pacific Ocean....By Edw. Freyhold. 1865–68. Scale 1 : 3,000,000.          **Q. 3**

Map of the Silurian Region. *See* Murchison (*Sir* R. I.).          72

**Marchand** (Étienne). *See* Fleurien (C. P. C.).          25

**\*Marey** (E. J.). Physiologie du mouvement. Le vol des Oiseaux. 8vo. *Paris*, 1890.          **Nf**

**\*Marine Biological Association** of the United Kingdom. Journal. N. S., Vol. 6, No. 4, Dec. N. S., Vol. 7, Nos. 1, 2, 3, 5. 8vo. *Plymouth*, 1903-6.          119

**Marion** (A. F.). *See* Saporta (*Le comte* G. de). Recherches sur les végétaux fossiles de Meximieux. 4to. *Lyon*, 1876.          **Q. 1**

*See* Saporta (*Le comte* G. de). Révision de la Flore Heersienne de Gelinden. (Extr.) 4to. *Bruxelles*, 1878.          **Nc**

*See* Saporta (*Le comte* G. de). L'évolution du règne végétal. Les cryptogames. 8vo. *Paris*, 1881.          57

**\*Marshall** (A. Milnes). The Frog : an Introduction to Anatomy and Histology. 8vo. *Manchester*, 1882.          106

**Marshall** (W.). Minutes of Agriculture, made on a Farm...near Croydon, Surry, &c. 4to. *London*, 1778. **Nf**

**Marshall** (?). A review of the Reports to the Board of Agriculture; from the Northern Department of England. 8vo. *York*, 1818. **49**

**Marshall** (Wm. E.). A Phrenologist amongst the Todas. 8vo. *London*, 1873. **124**

**Marsham** (Thomas). Coleoptera Britannica. Vol. 1. 8vo. *Londini* (1802). **112**

**\*Martin** (H. N.). *See* Huxley (T. H.). A course of elementary instruction in Practical Biology. 8vo. *London*, 1875 and 1881. **106**

**Martin** (W. C. L.). The history of the Dog. 8vo. *London*, 1845. **118**
The history of the Horse. 8vo. *London*, 1845. **118**

**Martin-Saint-Ange** (G. J.). Mémoire sur l'organisation des Cirripèdes. 4to. *Paris*, 1835. **Ne**

**Martins** (Charles). *See* Lamarck. Philosophie zoologique. Nouv. éd. 2 tomes. 8vo. *Paris*, 1873. **112**

**Martius** (C. F. Phil. von). *See* Spix (J. B. von). Travels in Brazil... 1817–20. 2 vols. 8vo. *London*, 1824. **15**

**Marx** (Karl). Das Kapital. 1er Bd. Buch 1. 2te Aufl. 8vo. *Hamburg*, 1872. **28**

**Marx** (K. F. H.). *See* Blumenbach (J. F.). The Anthropological Treatises of J. F. B. (Memoir by Marx.) Transl. 8vo. *London*, 1865. **114**

**Masaryk** (Thomas Garrigue). Der Selbstmord als sociale Massenerscheinung der modernen Civilisation. 8vo. *Wien*, 1881. **12**

**Maskelyne** (Nevil). Tables requisite to be used with the Nautical Ephemeris, &c. Ed. by N. M., Astronomer Royal. 3rd ed. 8vo. *London*, 1802. **14**

**Masters** (Maxwell T.). Vegetable Teratology. (Ray Soc. Publ.) 8vo. *London*, 1869. **17**

**Mathews** (Wm.). The Flora of Algeria. 8vo. *London*, 1880. **59**

**\*Maton** (Wm. G.). *See* Linnæus (Carl). A general view of the writings of Linnæus, by R. Pulteney. 2nd ed., by Wm. G. M. 4to. *London*, 1805. **Nc**

**Matthes** (Benno). Betrachtungen über Wirbelthiere, deren Seelenleben und die Stellung derselben zum Menschen. 8vo. *Dresden*, 1861. **12**

**Matthew** (Patrick). On Naval Timber and Arboriculture. 8vo. *Edinburgh*, 1831. **49**

**Maudsley** (Henry). The Physiology and Pathology of Mind. 2nd ed., rev. 8vo. *London*, 1868. 22
Body and Mind. 8vo. *London*, 1870. 22
do. Enlarged ed. 8vo. *London*, 1873. 22
The Physiology of Mind. 8vo. *London*, 1876. 22

**Maupas** (E.). *See* Burmeister (H.). Histoire de la Création. Trad. 8vo. *Paris*, 1870. 97

**Maw** (George). *See* Hooker (J. D.). Journal of a Tour in Marocco, &c. 8vo. *London*, 1878. 89

**Mawe** (John). Travels in the gold and diamond districts of Brazil. New ed. 8vo. *London*, 1825. 15

***Maxwell** (J. Clerk). Matter and Motion. 8vo. *London*, 1882. 24

**Mazaroz** (J. P.). La Genèse des sociétés modernes. 8vo. *Paris*, 1877. 28

**Medlicott** (H. B.). A Manual of the Geology of India. 2 Parts. 2 vols. (and Map dated 1877). Compiled by H. B. M., and W. T. Blanford. 8vo. *Calcutta*, 1879. **Nf**

**Meehan** (Thomas). The native Flowers and Ferns of the United States. 2 vols. 4to. *Boston*, 1878–79. 44

**Meetkerke** (C. E.). The Guests of Flowers. 8vo. *London*, 1881. 62

**Meitzen** (Ernst). Bhawani. 8vo. *Köln*, 1872. 39

**Melia** (Pius), *D.D.* Hints and facts on the Origin of Man, &c. 8vo. *London*, 1872. 9

**Mémoires de l'acad. imp....de Lyon.** Classe des sciences (Nouv. Série). Tome 2ème. 8vo. *Lyon*, 1852. *See* Jordan (A.). 44

**Mémoires (Nouveaux) de la soc. imp. des naturalistes de Moscou.** Tome 13, Livr. 4, 5 ; Tome 14, Livr. 1. 4to. *Moscou*, 1874–79. 18

**Memoirs of the Geological Survey of Great Britain,** and of the Museum of Economic Geology in London. Vol. 1. 8vo. *London*, 1846. **Nf**

**Memorie della reale accademia delle scienze di Torino.** Serie 2a. Tom. 27–33. 4to. *Torino*, 1873–81. 18

**Meneghini** (G.). *See* Henfrey (A.). Botanical and Physiological Memoirs. Transl. 8vo. *London*, 1853. 17

**Merriam** (C. Hart). A review of the Birds of Connecticut. 8vo. *New Haven*, 1877. 102

**Metzger** (I.). Die Getreidearten und Wiesengräser. 8vo. *Heidelberg*, 1841. 60

**Meyen** (F. J. F.). Beiträge zur Zoologie, gesammelt auf einer Reise um die Erde, und W. Erichson's und H. Burmeister's Beschreibungen und Abbildungen der...Insekten. 4to. *Breslau*, 1834. (Extr.) **Ne**
*Neues System der Pflanzen-Physiologie. 3 Bde. 8vo. *Berlin*, 1837–39. **62**
Outlines of the Geography of Plants. Transl. by Margaret Johnston. (Ray Soc. Publ.) 8vo. *London*, 1846. **17**

**Meyer** (F. A. A.). Versuch einer vollständigen Naturgeschichte der Hausthiere, im Grundrisse. 8vo. *Göttingen*, 1792. **108**

*Michael** (Albert D.). British Oribatidæ. 2 vols. (Ray Soc. Publ.) 8vo. *London*, 1884–88. **17a**

**Michell** (*Rev.* John). Conjectures concerning the cause...of Earthquakes. 4to. *London*, 1760. **50**

**Miers** (John). Travels in Chile and La Plata. 2 vols. 8vo. *London*, 1826. **27**

**Miller** (Hugh). Foot-prints of the Creator. 8vo. *London*, 1849. **9**

**Miller** (Philip). The Gardener's Dictionary. Abridged. In 3 vols. Vols. 1 and 3. 3rd ed. 8vo. *London*, 1748. **49**

*Miller** (William A.). Elements of Chemistry. Part 2. 3rd ed. 8vo. *London*, 1864. **105**

**Milligan** (Ed.). *See* Celsus (A. C.). A. C. Celsi Medicinae Libri Octo. 8vo. *Edinburgi*, 1826. **94**

**Milne-Edwards** (H.). Introduction à la Zoologie générale. 1ère Partie. 8vo. *Paris*, 1851. **106**

*Milner** (*Rev.* Thomas). A descriptive Atlas of Astronomy, and of physical and political Geography....The Maps by Augustus Petermann. 4to. *London*, 1853. **Nc**

**Mind.** Nos. 1, 3–19. 8vo. *London*, 1876–80. **37**

*Minnesota Botanical Studies.** 2nd Ser., Parts 2, 5, 6; 3rd Ser., Part 1. 8vo. 1899–1903. **49**

**Miquel** (Fred. A. G.). Disquisitio geographico-botanica de Plantarum Regni Batavi distributione. 8vo. *Lugd.-Batav.*, 1837. **60**

**Mitchell** (S. Weir). Researches upon the venom of the Rattlesnake. 4to. *Washington*, 1861. **Na**

**Mivart** (St George). On the Appendicular Skeleton of the Primates. (Extr.) 4to. *London*, 1867. **74**
On the Genesis of Species. 8vo. *London*, 1871. **23**
do. 2nd ed. 8vo. *London*, 1871. **23**
Lessons in Elementary Anatomy. 8vo. *London*, 1873. **105**
Man and Apes. 8vo. *London*, 1873. **10**

**\*Moffat** (C. B.). *See* More (Alex. G.). Life and Letters of A. G. M.
8vo. *Dublin*, 1898. 123

**Moggridge** (J. Traherne). Harvesting Ants and Trap-door Spiders.
8vo. *London*, 1873. 102
Supplement...with specific descriptions of the Spiders, by the Rev. O.
Pickard-Cambridge. 8vo. *London*, 1874. 102

**Mohl** (Hugo von). Ueber den Bau und das Winden der Ranken und
Schlingpflanzen. 4to. *Tübingen*, 1827. 44
\*Vermischte Schriften botanischen Inhalts. 4to. *Tübingen*, 1845. Nc
Principles of the Anatomy and Physiology of the Vegetable Cell.
Transl. by A. Henfrey. 8vo. *London*, 1852. 122

**Mohl** (Jules). Vingt-sept ans d'histoire des études orientales. 2 tomes.
8vo. *Paris*, 1879–80. 16

**Mojsisovics von Mojsvár** (Edmund). Die Dolomit-Riffe von Südtirol
und Venetien. 8vo. *Wien*, 1879. 97
do. Geologische Uebersichtskarte. Blatt 1. Fol. *Wien*, 1878. **Q.** 3
Das Gebirge um Hallstatt. 1 Theil, 1, 2 Heft. Die Mollusken-Faunen
der Zlambach- und Hallstätter-Schichten. (Extr.) 4to. *Wien*, 1873–75.
72
Ueber die triadischen Pelecypoden-Gattungen Daonella und Halobia.
(Extr.) 4to. *Wien*, 1874. 72

**Molendo** (Ludwig). *See* Walther (A.). 118

**Moleschott** (Jac.). La circulation de la vie...Trad. par E. Cazelles.
2 tomes. 8vo. *Paris*, 1866. 11
Der Kreislauf des Lebens. 1er Bd. 5te Aufl. 8vo. *Mainz*, 1877. 105
Sull' Influenza della luce mista e cromatica. Ricerche...da J. M. e
S. Fubini. (Extr.) 8vo. *Torino*, 1879. 105

**Molina** (Juan Ignacio). Compendio de la Historia Geografica...del
Reyno de Chile, escrito en Italiano. 1a Parte...Trad. en Español
por Domingo Joseph. 4to. *Madrid*, 1788. 24
Compendio de la Historia civil del Reyno de Chile...Parte segunda.
Trad. al Español...por Nicolas de la Cruz y Bahamonde. 4to.
*Madrid*, 1795. 24

**\*Molisch** (Hans). *See* Wiesner und seine Schule. Festschrift. 8vo.
*Wien*, 1903. 35

**Moll** (L.). La connaissance générale du Bœuf...Publiée...sous la direction
de L. Moll et Eug. Gayot. Avec un Atlas. 2 vols. 8vo. *Paris*,
1860. 108

**Moniez** (R.). Mémoires sur les Cestodes. 1ère Partie. (Extr.) 4to.
*Paris*, 1881. Ne

**Moore** (David). Contributions towards a Cybele Hibernica. By D. M.,
and A. G. More. 8vo. *Dublin*, 1866. 59

**Moore** (Frederic). Descriptions of New Indian Lepidopterous Insects... Rhopalocera, by W. C. Hewitson...Heterocera by Fr. Moore. 4to. *Calcutta*, 1879. **74**

**Moore** (George). The first Man and his place in Creation. 8vo. *London*, 1866. **124**

**Moore** (John). Columbarium : or the Pigeon-house. 8vo. *London*, 1735. [Included with J. M. Eaton's Treatise on Pigeons. 8vo. *London*, 1852.] **127**
do. [8vo. *London*, 1858.] **127**

**\*Moore** (Norman), *M.D.* The Book of the foundation of St Bartholomew's Church in London....Ed. from the original Manuscript. 8vo. 1886. **24**
\*The distribution and duration of Visceral new Growths. 8vo. *Edinburgh*, 1889. [2 copies.] **92**
\*Pathological Anatomy of Diseases. (Student's Guide Series.) 8vo. *London*, 1889. **104**
\*The Harveian Oration delivered...Oct. 18, 1901. 8vo. *London*, 1901. **22**
\*History of Medicine in Ireland. *See* Saint Bartholomew's Hospital Reports. **92**

**Moore** (Wm. D.). *See* Donders (F. C.). **104**

**Moquin-Tandon** (A.). Éléments de Tératologie végétale, &c. 8vo. *Paris*, 1841. **57**

**More** (Alex. G.). *See* Moore (D.). Contributions towards a Cybele Hibernica. 8vo. *Dublin*, 1866. **59**
\*Life and Letters of A. G. M....Ed. by C. B. Moffat. With a Preface by Frances M. More. 8vo. *Dublin*, 1898. **123**

**Morgan** (Lewis H.). The American Beaver and his works. 8vo. *Philadelphia*, 1868. **118**
Systems of Consanguinity and Affinity of the Human Family. 4to. *Washington*, 1871. **72**

**Morren** (Édouard). *See* Actes du Congrès de botanique horticole réuni à Bruxelles. 8vo. *Liége*, 1877. **59**

**Morris** (John). A Catalogue of British Fossils. 2nd ed. 8vo. *London*, 1854. **107**

**Morton** (Samuel G.). Types of Mankind. Illustrated by selections from the inedited Papers of S. G. M. (and others), by J. C. Nott, and Geo. R. Gliddon. 8vo. *Philadelphia*, 1854. **114**

**Moseley** (H. N.). On the structure and development of *Peripatus capensis*. (Extr.) 4to. *London* [Read 1874]. **Ne**
Oregon. 8vo. *London*, 1878. **89**
Notes by a Naturalist on the "Challenger"...1872–76. 8vo. *London*, 1879. **9**

**Mosso** (A.). Ueber den Kreislauf des Blutes im menschlichen Gehirn. 8vo. *Leipzig*, 1881. 115
*La Peur. Etude psycho-physiologique. Trad. par Félix Hément. 8vo. *Paris*, 1886. 22

**Moubray** (Bonington). A practical treatise on breeding...Poultry, Pigeons, and Rabbits. 7th ed. 8vo. *London*, 1834. 127

**Müller** (Aug.). Ueber die erste Entstehung organischer Wesen und deren Spaltung in Arten. 3te Aufl. 8vo. *Berlin*, 1881. 23

**Müller** (*Baron* Ferd. von). Fragmenta Phytographiæ Australiæ. Vol. 7. 8vo. *Melbourne*, 1869–71. 62
*Select extra-tropical Plants...for...Naturalisation. 7th ed. 8vo. *Melbourne*, 1888. 62

**Müller** (Friedrich). Reise der österreichischen Fregatte Novara...1857–59. Anthropologischer Theil, 3te Abth., Ethnographie. 4to. *Wien*, 1868. 67
Allgemeine Ethnographie. 8vo. *Wien*, 1873. 114

**Müller** (Fritz). Für Darwin. 8vo. *Leipzig*, 1864. 23

**Müller** (Hermann). Die Befruchtung der Blumen durch Insekten und die gegenseitigen Anpassungen beider. 8vo. *Leipzig*, 1873. 57
Alpenblumen, ihre Befruchtung durch Insekten und ihre Anpassungen an dieselben. 8vo. *Leipzig*, 1881. 57
*See Krause (E.). Hermann Müller von Lippstadt. 8vo. *Lippstadt*, 1884. 113

**Müller** (Johannes). Elements of Physiology. Transl. by Wm. Baly. 2 vols. 8vo. *London*, 1838–42. 125
Elements of Physiology. Suppl. to the 2nd vol. *See* Baly (W.) and Kirkes (W. S.). 125
Ueber die Gattungen der Seeigellarven. 4to. *Berlin*, 1855. Ne
*See* Du Bois-Reymond (E.). 8

**Munk** (Hermann). Die elektrischen und Bewegungs-Erscheinungen am Blatte der Dionaea muscipula. Von H. M., mit der anatomischen Untersuchungen des Dionaea-Blattes, von F. Kurtz. (Extr.) 8vo. *Leipzig*, 1876. 122

**Murchison** (Charles). *See* Falconer (Hugh). Palæontological Memoirs, &c. 2 vols. 8vo. *London*, 1868. 126

**Murchison** (*Sir* Roderick I.), *Bart.* The Silurian Region and adjacent Counties of England and Wales Geologically Illustrated...during the years 1831–38 (Map in 3 Sections). 72
On the Silurian System of Rocks. (Extr.) 8vo. *London*, 1835. [Philos. Tracts, ii. 16ª.] 11
The Silurian System. In two Parts. (3 vols.) 4to. *London*, 1839. 72
The Silurian System. (Extr.) 8vo. *Edinburgh*, 1841. 97

**Murphy** (Joseph J.). Habit and Intelligence. 2 vols. 8vo. *London*, 1869. **89**
Habit and Intelligence. 2nd ed. 8vo. *London*, 1879. **89**

**Murray** (Alex.). *See* Canada. Geological Survey. Plans of...Lakes and Rivers. 4to. *Toronto*, 1857. **74**

**Murray** (Andrew). The geographical distribution of Mammals. 4to. *London*, 1866. [2 copies.] **75**
*See* Journal (The) of Travel, &c. Vol. I. 8vo. *London*, 1868-69. **9**

**Murray** (Lindley). An English Grammar. In 2 vols. 5th ed. improved. 8vo. *York*, 1824. **16**

**Museum of Practical Geology, &c.** Records of the School of Mines... Vol. 1. Part 2. 8vo. *London*, 1853. **Ng**

**Nägeli** (Carl). Botanische Mittheilungen. 8vo. *München*, 1866. *Imperfect*. **44**
On vegetable cells. Transl. by A. Henfrey. (Ray Soc. Publ.—Reports and Papers on Botany.) 8vo. *London*, 1846. **17**
*Pflanzenphysiologische Untersuchungen. Von C. N., und Carl Cramer. 1-4 Hefte. 4to. *Zürich*, 1855-58. **Nc**
*Mechanisch-physiologische Theorie der Abstammungslehre. 8vo. *München*, 1884. **57**
Die niederen Pilze in ihren Beziehungen zu den Infectionskrankheiten und der Gesundheitspflege. 8vo. *München*, 1877. **55**
Theorie der Gärung. 8vo. *München*, 1879. **57**

**Naples.** Fauna und Flora des Golfes von Neapel. *See* C. Chun, C. Emery, A. Dohrn, Graf zu Solms-Laubach. **Na**

**Nash** (Wallis). Oregon: there and back in 1877. 8vo. *London*, 1878. **25**

**Nathusius** (Hermann von). Vorstudien für Geschichte und Zucht der Hausthiere zunächst am Schweineschädel. Mit einem Atlas. (Obl.) 8vo. *Berlin*, 1864. **75 & Q.1**
Vorträge über Viehzucht und Rassenkenntniss. Th. 1-3. 8vo. *Berlin*, 1872-80. **108**

**Natural History.** A Vol. of Plates. Publ. by Whittaker, 1824-26. 8vo. *London*. **26**

**Natural History Review.** Nos. 1-20. 8vo. *London*, 1861-65. **116**

**Natural Science, Religious Creeds, &c.** By the Author of 'The Divine Footsteps in Human History.' [By Daniel Reid.] Part 1. 8vo. *Edinburgh*, 1870. **14**

**Naudin** (Ch.). Nouvelles recherches sur l'hybridité dans les végétaux. (Extr.) 4to. *Paris*, 1862. **Nc**

**Nautical Almanack.** *See* Maskelyne (N.). **14**

**Nebraska.** General Survey. *See* R. Pound and Fr. E. Clements. 2nd ed. 8vo. *Lincoln, Neb.*, 1900. **59**

**Neill** (Pat.). *See* Journal of a Horticultural Tour through...Flanders, &c. 8vo. *Edinburgh*, 1823. **61**
*See* Daubuisson (J. F.). **107**

**Nelson** (Richard J.). On the Geology of the Bermudas. *See* Trans. Geol. Soc. of London, 2nd Ser., Vol. 5, Part 1. 4to. *London* [Read 1834]. **Nc**

**Netter** (A.). De l'Intuition dans les Découvertes et Inventions. 8vo. *Strasbourg*, 1879. **40**

**Neuman and Baretti's Dictionary of the Spanish and English Languages.** 5th ed. Vol. 1. *See* Dictionary. 8vo. *London*, 1831. **14**

**Neumayr** (M.). Die Congerein- und Paludinenschichten Slavoniens und deren Faunen. Von M. N., und C. M. Paul. (Extr.) 4to. *Wien*, 1875. **Na**
Zur Kenntniss der Fauna des untersten Lias in den Nordalpen. (Extr.) 4to. *Wien*, 1879. **Na**

**Neumeister** (Gottlob). Das Ganze der Taubenzucht. Obl. *Weimar*, 1837. **127**

**New (The) Truth and the Old Faith.** By a Scientific Layman. 8vo. *London*, 1880. **22**

**New York.** 18th Report of the Commissioners of Fisheries of the State of New York...Feb. 10, 1890. 8vo. *Albany*, 1890. **106**

**Newberry** (J. S.). The structure and relations of Dinichthys, &c. (Extr.) 8vo. *Columbus*, 1875. **97**

**\*Newell** (Jane H.). Outlines of Lessons in Botany 2 Parts. 8vo. *Boston, U.S.A.*, 1889–92. **62**

**Newton** (Alfred). Zoology. (Manuals of Element. Science.) 8vo. *London*, 1874. **96**

**Nicholson** (Edward). Indian Snakes. An elementary treatise on Ophiology. 2nd ed. 8vo. *Madras*, 1874. **106**

**\*Niemeyer** (Felix von). A Text-book of practical Medicine. Transl. from the 8th German ed. by G. H. Humphreys and C. E. Hackley. Rev. ed. 2 vols. 8vo. *London*, 1871. **94**

**Nilsson** (Sven). The Primitive Inhabitants of Scandinavia. 3rd ed. Ed., and with an Introd. by Sir J. Lubbock, Bart. 8vo. *London*, 1868. **124**

**Nitzsch** (Christian L.). Nitzsch's Pterylography. Transl. from the German. Ed. by P. L. Sclater. (Ray Soc. Publ.) Fol. *London*, 1867. **72**

**Noiré** (Ludwig). Die Welt als Entwicklung des Geistes. 8vo. *Leipzig*, 1874     **22**

Der Ursprung der Sprache. 8vo. *Mainz*, 1877.     **24**

Das Werkzeug und seine Bedeutung für die Entwickelungsgeschichte der Menschheit. 8vo. *Mainz*, 1880.     **23**

**Nordenskiöld** (A. E.). The Voyage of the Vega. Transl. by Alex. Leslie. In 2 vols. Vol. 2. 8vo. *London*, 1881.     **15**

**Norman** (*Rev.* A. M.). *See* Bowerbank (J. S.). A Monograph of the British Spongiadæ, Vol. 4. 8vo. *London*, 1882.     **17**

**Norton** (Charles Eliot). *See* Wright (Chauncey). Philosophical Discussions. 8vo. *New York*, 1877.     **40**

**Nott** (J. C.). *See* Morton (S. G.). Types of Mankind. 8vo. *Philadelphia*, 1854.     **114**

**Notter** (Fridr.). *See* Hofacker (J. D.).     **108**

***Nusbaum** (Joseph). Recherches sur l'organogénèse des Hirundinées (*Clepsine complanata Sav.*). (Extr.) 8vo. *Paris*, 1886.     **106**

L'Embryologie de Mysis Chameleo (Thompson). (Extr.) 8vo. *Varsovie*.     **106**

**Odart** (*Le Cte*). Ampélographie universelle ou Traité des Cépages. 2e éd. 8vo. *Paris*, 1849.     **59**

***Oels** (Walter). Pflanzenphysiologische Versuche. 8vo. *Braunschweig*, 1893.     **59**

**Ogilby** (W.). Observations on the opposable Power of the Thumb in certain Mammals, &c. (Extr.) 8vo. *London*, 1837. [Philos. Tracts, i. 15.]     **11**

**Ogle** (John W.). The Harveian Oration, 1880. 8vo. *London*, 1881.     **104**

**Ogle** (W.). *See* Kerner (A.). Flowers and their unbidden guests. 8vo. *London*, 1878.     **59**

*See* Aristotle.     **Nf**

***Oliver** (Joseph W.). Elementary Botany. 3rd ed. 8vo. *London*, 1892.     **62**

**O'Neill** (T. Warren). The Refutation of Darwinism. 8vo. *Philadelphia*, 1880.     **22**

**Ontario.** Annual Report of the Commissioner of Agriculture, &c....for 1872. 8vo. *Toronto*, 1873.     **49**

**Oppert** (Gustav). On the Classification of Languages. 8vo. *Madras*, 1879.     **24**

**Ord** (William Miller). On the influence of Colloids upon crystalline form and cohesion. 8vo. *London*, 1879.     **93**

**Ordinaire** (C. N.). Histoire naturelle des Volcans. 8vo. *Paris*, An x (1802).     **107**

**Ormathwaite** (*Lord*).  Astronomy and Geology compared.  8vo.
  *London*, 1872.  **10**

**Orton** (James).  The Andes and the Amazon.  8vo. *New York*, 1870.  **25**
  do.  3rd ed.  8vo. *New York*, 1876.  **25**

**Osborne** (J.).  The Horsebreeders' Handbook.  Ed. by J. O. ("Beacon")·
  8vo. *London*, 1881.  **118**

**Our Blood Relations;**  or, the Darwinian Theory.  8vo. *London*
  1872.  **9'**

**Outlines of History.**  (Lardner's Cab. Cyclop.)  8vo. *London*, 1830.
  **128**

**Ovington** (J.).  A voyage to Suratt, in the year 1689, &c.  8vo. *London*
  1696.  **26**

**Owen** (John Pickard), *pseud.*  The Fair Haven.  By the late J. P. O.
  Ed. by W. B. Owen.  8vo. *London*, 1873.  (By S. Butler.)  **24**

**Owen** (*Sir* Richard).  Fossil Mammalia (Zoology of...H.M.S. Beagle·
  Part 1).  4to. *London*, 1840.  [2 copies.]  **67**
  Description of the Skeleton of an extinct gigantic Sloth, *Mylodon
  robustus*, Owen.  4to. *London*, 1842.  **Na**
  Description of certain Fossil Crania, discovered by A. G. Bain, in...
  S.E. Africa...(Dicynodon).  (Extr.)  4to. *London*, 1845.  **Nc**
  A history of British Fossil Mammals, and Birds.  8vo. *London*, 1846.
  **107**
  The nature of Limbs.  8vo. *London*, 1849.  **94**
  On Parthenogenesis.  8vo. *London*, 1849.  [2 copies.]  **94**
  Lectures on the Comparative Anatomy and Physiology of the Inverte-
  brate Animals.  2nd ed.  8vo. *London*, 1855.  **115**
  Palæontology.  8vo. *Edinburgh*, 1860.  **126**
  do.  2nd ed.  8vo. *Edinburgh*, 1861.  **126**
  *See* Hunter (John).  Essays and Observations on Natural History, &c.
  2 vols.  8vo. *London*, 1861.  **13**
  On the Anatomy of the Vertebrates.  3 vols.  8vo. *London*, 1866–68.
  **115**

**\*P.** (G. H.), and J. B. P.  *See* "Authors and Publishers."  8vo. *London*,
  1897.  **24**

**Packard** (A. S.).  A Guide to the Study of Insects.  Parts 1–5.  8vo.
  *Salem*, 1868–69.  [Incomplete.]  **102**

**Packard** (A. S.), Jr.  *See* United States Entomological Commission.
  2nd Report, 1878–79.  8vo. *Washington*, 1880.  **102**
  Insects injurious to Forest and Shade Trees.  *See* U. S. Entomological
  Commission.  Bulletin, No. 7.  8vo. *Washington*, 1881.  **102**

**Page** (David).  Man where, whence, and whither.  8vo. *Edinburgh*,
  1867.  **9**

D.  5

**Paget** (*Sir* James). Lectures on Surgical Pathology. Vol. 1. 8vo.
  *London*, 1853.                                                   93
  do. 3rd ed. Rev. and ed. by Wm. Turner. 8vo. *London*, 1870.   93

**Palæontographical Society Publications.** 1848–81. 4to. *London*.
                                                                78, 79

**Palm** (Ludwig H.). Ueber das Winden der Pflanzen. (Dissertation.)
  8vo. *Tübingen*, 1827.                                            62

**Paolucci** (Luigi). Il canto degli Uccelli. 8vo. *Milano*, 1878. [2 copies.]
                                                                   106

**Paris** (J. A.). Pharmacologia. 6th ed. 2 vols. 8vo. *London*, 1825. 93
  The elements of Medical Chemistry. 8vo. *London*, 1825.        104

**Parish** (*Sir* Woodbine). Account of a Voyage to explore the River
  Negro [in 1782–83, by Don Basilio Villarino]. (Extr.) 8vo. *London*.
  [Philos. Tracts, ii. 7.]                                         11
  On the Southern Affluents of the River Amazons...communicated by
  W. P. 8vo. *London*, 1835. (Extr.) [Philos. Tracts, ii. 9.]     11

**Parker** (W. Kitchen). A Monograph on the Structure and Development
  of the Shoulder-girdle and Sternum in the Vertebrata. (Ray Soc.
  Publ.) Fol. *London*, 1868.                                      72

**\*Parkes** (Edmund A.). A manual of practical Hygiene. 4th ed. 8vo.
  *London*, 1873.                                                  92

**Parkinson** (James). An introduction to the study of Fossil Organic
  Remains. 8vo. *London*, 1822.                                   107

**Patterson** (H. S.). *See* Morton (S. G.).                      114

**Pauchon** (A.). Recherches sur le rôle de la Lumière dans la Germi-
  nation. 8vo. *Paris*, 1880.                                      57

**Paul** (C. M.). *See* Neumayr (M.).                             Na

**\*Peckham** (George W. and Elizabeth G.). On the instincts and habits
  of the solitary Wasps. 8vo. *Madison, Wis.*, 1898.              95

**Pelzeln** (Aug. von). *See* Holub (Emil).                      117

**Pennant** (Thomas). History of Quadrupeds. 3rd ed. 2 vols. 4to.
  *London*, 1793.                                                  75

**\*Pennsylvania.** Publications of the University. Contributions from
  the Botanical Laboratory. Vol. 2, No. 2. 8vo. *Philadelphia*, 1901.
                                                                   119

**Pentland** (J. B.). On the General Outline and Physical Configuration
  of the Bolivian Andes. (Extr.—Read 1835.) 8vo. *London*, 1835.
  [Philos. Tracts, ii. 8.]                                         11

**Pernety** (A. J.). Journal historique d'un voyage fait aux Îles Malouïnes
  en 1763–64. 2 tomes. 8vo. *Berlin*, 1769.                       10

**Perrier** (Edmond). Les colonies animales et la formation des organismes. 8vo. *Paris*, 1881. **Nf**

**Persoon** (C. H.). *See* Linnæus (C.). Systema Vegetabilium. Ed. 15a. 8vo. *Gottingae*, 1797. **61**
Synopsis Plantarum. Pars 1, 2. 8vo. *Parisiis Lutetiorum*, 1805–7. **62**

**Petermann** (Augustus). *See* Milner (*Rev.* Th.). A descriptive Atlas of Astronomy, &c. 4to. *London*, 1853. **Nc**

**Pettigrew** (James Bell). On the physiology of Wings. (Extr.) 4to. *Edinburgh*, 1871. **Ne**
The Physiology of the Circulation. 8vo. *Edinburgh*, 1873. **125**

**\*Pflüger** (E.). Ueber die physiologische Verbrennung in den lebendigen Organismen. (Extr.) 8vo. *Bonn* (1875). **105**

**Phillips** (John). A treatise on Geology. 2 vols. (Lardner's Cab. Cyclop.) 8vo. *London*, 1837–39. **128**
The Rivers, Mountains and Sea-Coast of Yorkshire. 2nd ed. 8vo. *London*, 1855. **24**
Life on the Earth. 8vo. *Cambridge*, 1860. **39**
Vesuvius. 8vo. *Oxford*, 1869. **117**
Geology of Oxford and the Valley of the Thames. 8vo. *Oxford*, 1871. **97**

**Phillips** (Wm.). An elementary introduction to…Mineralogy. 3rd ed. 8vo. *London*, 1823. 4th ed.…by R. Allan. 8vo. *London*, 1837. **117**
*See* Conybeare (*Rev.* W. D.). **107**

**Physicus**, *pseud.* A candid examination of Theism. *See* G. J. Romanes. **12**

**Pickard-Cambridge** (*Rev.* O.). *See* Moggridge (J. Traherne). Supplement to 'Harvesting Ants,' &c. 8vo. *London*, 1874. **102**

**Pickering** (Charles). The Races of Man. New ed. (Bohn's illustr. Library.) 8vo. *London*, 1850. **124**

**Pictet** (François Jules). Traité élémentaire de Paléontologie. Tomes 1, 3. 8vo. *Genève*, 1844–45. **126**
Traité de Paléontologie. 2de éd. Tomes 1–4. 8vo. *Paris*, 1853–57. **126**
Traité de Paléontologie. Atlas. 4to. *Paris*, 1853–57. **74**
*See* Soret (J. Louis). **113**

**Piderit** (Theodor). Wissenschaftliches System der Mimik und Physiognomik. 8vo. *Detmold*, 1867. **39**

**Pistor** (E. M. W.). Das Ganze der Feld- und Hoftaubenzucht. 8vo. *Hanau*, 1831. **127**

**\*Pizzetta** (J.). Galerie des Naturalistes. 8vo. *Paris*, 1891. **9**

**Planck** (K. Ch.). Seele und Geist. 8vo. *Leipzig*, 1871. **22**
Wahrheit und Flachheit des Darwinismus. 8vo. *Nördlingen*, 1872. **39**

**Playfair** (J.). Illustrations of the Huttonian Theory. 8vo. *Edinburgh*, 1802. 113

**Playfair** (Lyon). *See* Liebig (Justus). Organic chemistry. 8vo. *London*, 1840. 24

**Pompper** (Hermann). Die Säugethiere, Vögel und Amphibien nach ihrer geographischen Verbreitung tabellarisch zusammengestellt. 4to. *Leipzig*, 1841. **Na**

**Population.** A Theory of Population, deduced from the General Law of Animal Fertility. *See* The Westminster Review, N. S., April, 1852. 119

**P\*\*\*\*** (M. F.). *See* Du Fuchsia. 8vo. *Paris*, 1844. 62

**Portanova** (Gennaro). Errori e delirii del Darwinismo. 8vo. *Napoli*, 1872. 10

**\*Portheim** (Leopold R. v.). *See* Wiesner und seine Schule. Festschrift. 8vo. *Wien*, 1903. 35

**Posnett** (Hutcheson M.). The historical method in Ethics, Jurisprudence, and Political Economy. 8vo. *London*, 1882. 28

**Pouchet** (Georges). The Plurality of the Human Race. Transl. and ed. (from the 2nd ed.), by Hugh J. C. Bevan. 8vo. *London*, 1864. 12

**\*Poulton** (E. B.) on Huxley. The Quarterly Review. No. 385. Jan., 1901. 113

**Poultry Chronicle.** Vols. 1–3. 4to. *London*, 1854–55. 21

**\*Pound** (Roscoe). The Phytogeography of Nebraska. i. General Survey. By R. P., and F. E. Clements. 2nd ed. 8vo. *Lincoln, Neb.*, 1900. 59

**Pourtalès** (L. F. de). Deep-Sea Corals. (Extr.) 4to. *Cambridge, Mass.*, 1871. 74
*See* Agassiz (Alex.). Echini, Crinoids, and Corals. (Extr.) 4to. *Cambridge, Mass.*, 1874. 74

**Powell** (J. W.). Introduction to the study of Indian Languages. 2nd ed. 4to. *Washington*, 1880. [2 copies.] 74

**Pozzi** (Samuel). Du Crane. (Extr.) 8vo. *(Paris)*, 1879. **Ne**

**Preyer** (W.). Die Blausäure. Physiologisch Untersucht. 8vo. *Bonn*, 1870. 115
Die Blutkrystalle. 8vo. *Jena*, 1871. 115
Das myophysische Gesetz. 8vo. *Jena*, 1874. 115
Naturwissenschaftliche Thatsachen und Probleme. Populäre Vorträge. 8vo. *Berlin*, 1880. 23
Die Seele des Kindes. 8vo. *Leipzig*, 1882. 47

**Price** (John). Old Price's Remains. 8vo. *London*, 1863–64. 113

**Prichard** (James Cowles). Researches into the Physical History of Man. 5 vols. 3rd ed. ; (5 = 1st ed.). 8vo. *London*, 1836–47. **114**
do. Vols. 1, 2. 4th ed. 8vo. *London*, 1851. **114**

**Principles** of Organic Life. 8vo. *London*, 1868. **125**

**Proctor** (Richard A.). Pleasant ways in Science. 8vo. *London*, 1879. **10**

**Psychological Inquiries**: in a series of Essays. (By *Sir* Benj. C. Brodie.) 8vo. *London*, 1854. **12**

**Public Libraries** in the United States of America. Special Report. Part 1. (Ed. by S. R. Warren and S. N. Clark.) 8vo. *Washington*, 1876. **16**

**Pugin** (A. Welby). Contrasts. 4to. *London*, 1841. **8**

**\*Pulteney** (Richard). *See* Linnæus (Carl). A general view of the writings of Linnæus. 2nd ed., by W. G. Maton. 4to. *London*, 1805. **Nc**

**Putsche** (Carl W. E.). Taubenkatechismus. 8vo. *Leipzig*, 1830. **127**

**Puvis** (M. A.). De la Dégénération et de l'Extinction des variétés de Végétaux. 8vo. *Paris*, 1837. **60**

**Quadri** (Achille). Note alla Teoria Darwiniana. 8vo. *Bologna*, 1869. **39**

**\*Quarterly Review.** No. 385. Jan. 1901. **113**

**Quatrefages.** *See* Quatrefages de Bréau.

**Quatrefages de Bréau** (J. L. A.). Souvenirs d'un Naturaliste. 2 tomes. 8vo. *Paris*, 1854. **11**
Physiologie comparée. Les Métamorphoses. (Extr.) 8vo. *Paris*, 1855. **125**
Études sur les maladies actuelles du Ver à soie. (Extr.) 4to. *Paris*, 1859. **Ne**
Nouvelles Recherches faites en 1859 sur les maladies actuelles du Ver à soie. 4to. *Paris*, 1860. **Ne**
Unité de l'Espèce humaine. (Extr.) 8vo. *Paris*, 1861. **40**
Métamorphoses de l'homme et des animaux. 8vo. *Paris*, 1862. **11**
Histoire naturelle des Annelés marins et d'eau douce. Tomes 1, 2 [with Plates]. (Nouv. Suites à Buffon.) 8vo. *Paris*, 1865. **112**
M. de Q., en présentant à l'Académie un ouvrage de M. Vogt : Mémoire sur les Microcéphales ou Hommes-Singes. (Extr.) 4to. *Paris*, 1867. **Nc**
Charles Darwin et ses précurseurs français. 8vo. *Paris*, 1870. **39**
A la Mémoire de J. L. A. de Q. de B. 10 Février 1810—12 Janvier 1902. 4to. **Q. 1**

**Quetelet** (A.). Sur l'Homme et le développement de ses Facultés. 2 tomes. 8vo. *Paris*, 1835. **39**

**Rabbit-Book** (The) for the many. 8vo. *London*, n. d.     **128**

**Rabenhorst.** *See* Dictionary (German). 24mo. *London*, 1829.    **98**

**Radcliffe** (Charles B.). Dynamics of Nerve and Muscle. 8vo. *London*, 1871.    **92**

**Radenhausen** (C.). Osiris. Weltgesetze in der Erdgeschichte. 3 Bde. 8vo. *Hamburg*, 1875–76.    **40**

**Ram** (James). The Philosophy of War. 8vo. *London*, 1878.    **24**

**Rames** (J. B.). La Création d'après la Géologie et la Philosophie naturelle. (1ère Partie.) 8vo. *Paris*, 1869.    **107**

**Ramsay** (Andrew C.). A descriptive Catalogue of the Rock Specimens in the Museum of Practical Geology. By A. C. R., H. W. Bristow, and H. Bauerman. 8vo. *London*, 1858.    **107**
The Physical Geology and Geography of Great Britain. 8vo. *London*, 1863. 3rd ed. 8vo. *London*, 1872. 5th ed. 8vo. *London*, 1878.    **107**

**Rang** (Sander). Manuel de l'histoire naturelle des Mollusques et leurs coquilles. 12mo. *Paris*, 1829.    **118**

**Ranke** (Johannes). Grundzüge der Physiologie des Menschen. 3te Aufl. 8vo. *Leipzig*, 1875.    **125**

**Ray** (John). The Wisdom of God manifested in the Works of the Creation. In 2 Parts. 8vo. *London*, 1692.    **128**
Memorials of J. R. Ed. by Edwin Lankester. (Ray Soc. Publ.) 8vo. *London*, 1846.    **17**

**Ray Society Publications.** Fol. & 8vo. *London*.    **72, 17, 17a**

**Reade** (T. Mellard). Chemical Denudation in relation to Geological Time. 8vo. *London*, 1879.    **97**

**Reade** (Winwood). The Martyrdom of Man. 8vo. *London*, 1872.    **12**
The African Sketch-Book. 2 vols. 8vo. *London*, 1873.    **26**

**Record of Zoological Literature.** *See* Zoological Record.    **51**

**Rée** (Paul). Der Ursprung der moralischen Empfindungen. 8vo. *Chemnitz*, 1877.    **24**

**Reeve** (Lovell). The land and freshwater Mollusks indigenous to, or naturalized in, the British Isles. 8vo. *London*, 1863.    **106**

**Reichenau** (Wilhelm von). Die Nester und Eier der Vögel. (Darwinistische Schriften, Nr. 9.) 8vo. *Leipzig*, 1880.    **41**

**Reid** (Daniel). *See* Natural Science, Religious Creeds, &c.    **14**

**Reinhardt** (J.). *See* Flower (W. H.). Recent Memoirs on the Cetacea. (Ray Soc. Publ.) Fol. *London*, 1866.    **72**

**\*Reinke** (Johannes). Untersuchungen über die Quellung einiger vegetabilischer Substanzen. 8vo. *Bonn*, 1879.    **61**

**Rendu** (Victor). L'Intelligence des Bêtes. 8vo. *Paris*, 1863. 11

**\*Rengade** (J.). La Création naturelle et les êtres vivants. 8vo. *Paris*, 1883. 44

**Rengger** (J. R.). Naturgeschichte der Säugethiere von Paraguay. 8vo. *Basel*, 1830. 118

**Reports and Papers on Botany.** (Ray Soc. Publ.) 8vo. *London*, 1846. *See* J. G. Zuccarini, Grisebach, C. Nägeli, and H. F. Link. 17

**Reports on the Progress of Zoology and Botany.** 1841–42. (Ray Soc. Publ.) 8vo. *London*, 1845. *See* Prince C. L. Bonaparte, A. Wagner, F. H. Troschel, W. F. Erichson, C. Th. v. Siebold, and H. F. Link. 17

**Reports on Zoology for 1843-44.** Transl. from the German by G. Busk, A. Tulk and A. H. Haliday. (Ray Soc. Publ.) 8vo. *London*, 1847. *See* W. F. Erichson, C. T. v. Siebold, F. H. Troschel, and A. Wagner. 17

**Retzius** (Gustaf). Anatomische Untersuchungen. I. i., Das Gehörlabyrinth der Knochenfische. 4to. *Stockholm*, 1872. Nb
Finska Kranier. Fol. *Stockholm*, 1878. Q. 1
Das Gehörorgan der Wirbelthiere. I. Fol. *Stockholm*, 1881. Q. 1

do. and **Key** (Axel). Studien in der Anatomie des Nervensystems und des Bindegewebes. 2 vols. Fol. *Stockholm*, 1875–76. Q. 3

**Reuss** (G. Ch.). Pflanzenblätter in Naturdruck. Plates 2, 21, 24, 26, 28, 41. Fol. *Stuttgart.* Nb

**\*Rhys** (Gweirydd ap). *See* Dictionary (Welsh-English). 8vo. *Carnarvon* (1866). 98

**Rialle** (Girard de). La Mythologie comparée. Tome 1er. 8vo. *Paris*, 1878. 124

**Ribeiro** (Carlos). Noticia de algumas estações e monumentos prehistoricos. (Com a traducção em francez.) 4to. *Lisboa*, 1878. Ne

**Ribot** (Th.). Heredity. From the French of Th. R. 8vo. *London*, 1875. 13
L'Hérédité psychologique. 2ème éd. 8vo. *Paris*, 1882. 13

**Richard** (Louis Claude). Démonstrations botaniques...Publiées par H. A. Duval (d'Alençon). 12mo. *Paris*, 1808. 62

**Richardson** (H. D.). Pigs. 8vo. *Dublin*, 1847. 118

**Richardson** (*Sir* John). On Aplodontia...constituted for the reception of the Sewellel, a burrowing animal, &c. (Extr.) 8vo. (*London*, 1829). [Philos. Tracts, i. 19.] 11
Zoological Remarks. ["North of the great Canada lakes."] 8vo. [Philos. Tracts, i. 21.] 11

Remarks on the Climate, &c....of the Hudson's Bay Countries. (Extr.)
8vo. *Edinburgh*, 1825. [Philos. Tracts, i. 20.] **11**
Fauna Boreali-Americana. Parts 1–3. 3 vols. 4to. *London*, 1829–36.
**75**
Report on North American Zoology. (Extr.) 8vo. *London*, 1836.
[Philos. Tracts, i. 18.] **11**
Fishes. 5 Parts. (The Zoology of the Voyage of H.M.S. Erebus and
Terror...1839–43.) 4to. *London*, 1844–48. **67**

**\*Richet** (Charles). Essai de Psychologie générale. 8vo. *Paris*, 1887.
**22**

**Riedel** (Wilhelm). Die vorzüglichst bekannten Feinde der Tauben.
8vo. *Ulm*, 1824. **127**
Die Taubenzucht. 8vo. *Ulm*, 1824. **127**

**Riley** (Charles V.). 3rd to 9th Annual Reports on the noxious, bene-
ficial, and other Insects of the State of Missouri. 8vo. *Jefferson City*,
1871–77. **102**
The Locust Plague in the United States. 8vo. *Chicago*, 1877. **102**
Report of the Entomologist (Dept. of Agriculture). Author's ed. Aug.
22, 1879. 8vo. *Washington*, 1879. **102**
The Cotton Worm. *See* U. S. Entomological Commission. Bulletin,
No. 3. 8vo. *Washington*, 1880. **102**
*See* United States Entomological Commission. 2nd Report, 1878–79.
8vo. *Washington*, 1880. **102**

**Riley** (James). Loss of the American Brig Commerce...Aug. 1815.
4to. *London*, 1817. **8**

**Rimpau** (Wilh.). Züchtung auf dem Gebiete der landwirthschaftlichen
Kulturpflanzen. 8vo. **62**

**Ritchie** (Archibald Tucker). The Creation. 5th ed. 8vo. *London*,
1874. **10**

**Rivero**, de. Memoria sobre el rico mineral de Pasco. (Extr.) 8vo.
Philos. Tracts, i. 26. *See* Adriasola. **11**

**Rivers** (Thomas). The Miniature Fruit Garden. 3rd ed. 4to. *London*,
1848. **59**

**\*Roberts** (George). Topography and Natural History of Lofthouse and
its neighbourhood. 2 vols. 8vo. *London*, and *Leeds*, 1882–85. **16**

**Robinet.** Manuel de l'éducateur de Vers à soie. 8vo. *Paris*, 1848. **95**

**\*Robinson** (B. L.). Flora of the Galapagos Islands. (Extr.) 8vo.
1902. **60**

**Roemer** (Ferd.). Lethaea geognostica...Hrsg. von einer Vereinigung von
Paläontologen. 1. Th. Lethaea palaeozoica von F. R. *Atlas*. 8vo.
*Stuttgart*, 1876. **Nd**

*Rolleston (George). Forms of Animal Life. 8vo. *Oxford*, 1870.
[2 copies.] 116
*See* Greenwell (W.). British Barrows. 8vo. *Oxford*, 1877. 114

*Rolph (W. H.). Biologische Probleme, zugleich als Versuch einer Rationellen Ethik. 8vo. *Leipzig*, 1892. 40

Romanes (George John). Observations on the Locomotor system of Echinodermata. By G. J. R., and J. C. Ewart. (Extr.) 4to. *London*, 1881. 74
Mental Evolution in Animals...With a posthumous Essay on Instinct, by Charles Darwin. 8vo. *London*, 1885. 47
*Mental Evolution in Man. 8vo. *London*, 1888. 28

[Romanes (G. J.).] A candid examination of Theism. By Physicus. 8vo. *London*, 1878. 12

Roscoe (E. S.). *See* Derby (Edw. H., 15th Earl of). 123

Rosenbusch (H.). Mikroskopische Physiographie der massigen Gesteine. 8vo. *Stuttgart*, 1877. 107

Ross (James). The Graft Theory of Disease. 8vo. *London*, 1872. 93

Rossi (D. C.). Le Darwinisme et les générations spontanées. 8vo. *Paris*, 1870. 10

Roux (Wilhelm). Der Kampf der Theile im Organismus. 8vo. *Leipzig*, 1881. 57

Royal Geographical Society of London. Journal. Vols. 1–15, 17–24. General Index to the second ten volumes. 8vo. *London*, 1830–54. 20
Proceedings. Vols. 1–6. 8vo. *London*, 1855–62. 20

Royal Horticultural Society of London. Journal. Vol. 2, Parts 1 and 2, 1847; Vol. 5, Parts 2 and 3, 1850; N. S., Vol. 1, 1866; Vol. 2, Part 1, 1868. 8vo. *London*. 119

Royal Irish Academy. Proceedings. Vol. 10, Part 2. 8vo. *Dublin*, 1868. 119
do. Ser. 2, Vol. 2, July, 1875. 8vo. *Dublin*, 1875. 119

Royal Society of Edinburgh. Proceedings. 8vo. Vol. 5, pp. 457–end; Vol. 6, pp. 1–172, 391–end; Vol. 7, pp. 1–527; Vol. 8; Vol. 9, pp. 1–202, 471–end; Vol. 10; Vol. 11, pp. 1–322. 1865–. 54
Transactions. 4to. Vol. 24, Part 2; Vol. 25, Part 2; Vols. 26, 27, 28, 29; Vol. 30, Part 1. 1865–. 54

Royal Society of London. Catalogue of Scientific Papers (1800–83). 12 vols. 4to. *London*, 1867–1902. 66
*Obituary Notices. (With Index to previous Obituary Notices.) 1900–1. 63
do. (Reprint.) 1900–4. 8vo. *London*. 63

74

Proceedings. Vols. 1–6 (incomplete); Vols. 7–71; Vols. 72–74
(incomplete); Vols. 77–.                                     52, 53, 63
*Reports of the Commission appointed…for the investigation of Mediter-
ranean Fever. Parts 1–7. 8vo. *London*, 1905–7.             63
*Reports to the Evolution Committee, 1 and 2. 8vo. *London*, 1902,
1905.                                                        63
*Reports to the Malaria Committee. 1–8 Series. 8vo. *London*, 1900–3.
63
*Reports of the Sleeping Sickness Commission. Nos. 1–7. 8vo.
*London*, 1903–5.                                            63
The Philosophical Transactions and Collections, to the end of the year
1700. Abridg'd, &c. In 3 vols. By John Lowthorp. 4to. *London*,
1705.                                                        71a
Philosophical Transactions. Vol. 66, Part 1, 1776; Vols. 78–177,
1788–1886; Vols. 178–198A and B, 1887–1906; Vols. 199–206A,
1902–. 4to. *London*.                                        68—76

**Royer** (*Mme.* Clémence). Origine de l'homme et des sociétés. 8vo.
*Paris*, 1870.                                               13

**Rudolphi** (Karl Asmund). Beyträge zur Anthropologie und allgemeinen
Naturgeschichte. 8vo. *Berlin*, 1812.                        114

**Rütimeyer** (L.). Beiträge zur Kenntniss der fossilen Pferde, &c.
(Abdr.) 8vo. *Basel*, 1863.                                  97
Die Grenzen der Thierwelt. 8vo. *Basel*, 1868.               39
do. (Translated extracts in MS., 83 pp.)                     39
Die Rinder der Tertiär-Epoche, &c. 2ter Th. (Extr.) 4to. *Zürich*,
1878.                                                        74
Die Fauna der Pfahlbauten der Schweiz. 4to. (*Basel*, 1861). **Ne**

**Sabatier** (Armand). Études sur le Cœur et la Circulation centrale dans
la série des Vertébrés. 4to. *Montpellier*, 1873.            75

**Sabine** (*Lt.-Col.* Edward). *See* Humboldt (A. von). Cosmos. Transl.
8vo. *London*, 1846–48.                                      26

**Sachs** (Julius). Lehrbuch der Botanik nach dem gegenwärtigen Stand
der Wissenschaft. 2te Aufl. 8vo. *Leipzig*, 1870.            58
do. 3te Aufl. 8vo. *Leipzig*, 1873.                          58
Traité de Botanique conforme à l'état présent de la science…Traduit de
l'allemand sur la 3e édition et annoté par Ph. Van Tieghem. 8vo.
*Paris*, 1874.                                               58

**Sageret.** Mémoire sur les Cucurbitacées. (Extr.) 8vo. *Paris*, 1826.
49
Pomologie physiologique. 8vo. *Paris*, 1830.                 49

*****Saint Bartholomew's Hospital** Reports. Ed. by J. Andrew and
T. Smith. Vol. 11. 8vo. *London*, 1875.                      92

**St Clair** (George). Darwinism and Design. 8vo. *London*, 1873.   10

**Saint-Hilaire** (Auguste de). Leçons de Botanique. 8vo. *Paris*, 1841.
122
Voyage aux sources du Rio de S. Francisco, &c. 2 tomes. 8vo. *Paris*,
1847–48. 9

**Saint-Hilaire** (Geoffroy). Principes de Philosophie zoologique. 8vo.
*Paris*, 1830. 112
Vie, travaux et doctrine scientifique d'Étienne G. Saint-Hilaire. Par son
fils Isidore G. Saint-Hilaire. 8vo. *Paris*, 1847. 11

**Saint-Hilaire** (Isidore G.). Histoire...des Anomalies de l'organisation
chez l'Homme et les Animaux. 3 tomes. Avec Atlas. 8vo. *Paris*,
1832–37. 39
Essais de Zoologie générale (et Planches). (Suites à Buffon.) 8vo.
*Paris*, 1841. 112
Histoire naturelle générale des règnes organiques. 3 tomes. 8vo. *Paris*,
1854–62. 40

**St John** (Charles). A tour in Sutherlandshire, &c. 2 vols. 8vo.
*London*, 1849. 25
*Sketches of the Wild Sports and Natural History of the Highlands.
Illustr. ed. 8vo. *London*, 1878. 16

**Salter** (J. W.). A descriptive Catalogue of all the genera and species
contained in the accompanying Chart of Fossil Crustacea...Arr. and
drawn by J. W. S., and H. Woodward. [And Extract from the
Geolog. Mag., Oct. 2, 1865.] 4to. *London* (1865). 97

**Salter** (John). The Chrysanthemum. 8vo. *London*, 1865. 62

**Samouelle** (George). The Entomologist's Useful Compendium. 8vo.
*London*, 1819. 112

**Sanderson** (J. B.). *See* Burdon-Sanderson.

**Sanderson** (*Sir* T. H.), *K.C.B.* *See* Derby (Edw. H., 15th Earl of).
123

**Saporta** (*Le Comte* G. de). Recherches sur les végétaux fossiles de
Meximieux. Par G. de S., et A. F. Marion. Précédées d'une intro-
duction stratigraphique par Albert Falsan. (Extr.) 4to. *Lyon*, 1876.
**Q. 1**
Sur le nouveau groupe paléozoïque des Dolérophyllées. (Extr.) 4to.
*Paris*, 1878. **Nc**
Révision de la Flore Heersienne de Gelinden....Par le comte G. de S.,
et A. F. Marion. (Extr.) 4to. *Bruxelles*, 1878. **Nc**
Observations sur la nature des végétaux réunis dans le groupe des
*Noeggerathia.* (Extr.) 4to. *Paris*, 1878. **Nc**
Le Monde des Plantes avant l'apparition de l'homme. 8vo. *Paris*, 1879.
56
L'évolution du règne végétal. Les cryptogames. Par G. de S., et
A. F. Marion. 8vo. *Paris*, 1881. 57

**\*Sargent** (Charles Sprague). *See* Gray (Asa). Scientific Papers of A. G.
2 vols. 1834–86. 8vo. *Boston*, 1889. 113

**Sarmiento de Gambóa** (Pedro). Viage al estrecho de Magallanes.
Por el Capitan P. S. de G. en los años de 1579 y 1580. 8vo. *Madrid*,
1768. 25

**Sartorius von Waltershausen** (W.). Untersuchungen über die
Klimate der Gegenwart und der Vorwelt. (Extr.) 4to. *Haarlem*,
1865. Ne

**Savage** (M. J.). The Religion of Evolution. 8vo. *Boston*, 1876. 12

**Schaaffhausen** (Hermann). Ueber den Zustand der wilden Völker.
(Archiv für Anthropologie, 1er Bd, 2tes Heft.) 4to. *Braunschweig*,
1866. Ne
Die Lehre Darwin's und die Anthropologie. (Archiv für Anthropologie,
3er Bd, 3tes u. 4tes Heft.) 4to. *Braunschweig*, 1869. Ne

**Schacht** (Hermann). The Microscope. Ed. by Fr. Currey. 2nd ed.
8vo. *London*, 1855. 62
Madeira und Teneriffe mit ihrer Vegetation. 8vo. *Berlin*, 1859. 15

**Scherzer** (Karl v.). Aus dem Natur- und Völkerleben im tropischen
Amerika. 8vo. *Leipzig*, 1864. 15
*See* Weisbach (A.). Reise der österr. Fregatte Novara. Anthrop. Theil,
2te, 3te Abth. 4to. *Wien*, 1867–68. 67
La Province de Smyrne. Trad. de l'allemand par F. Silas. 8vo.
*Vienne*, 1873. 15

**Schiff** (Maurice). Leçons sur la physiologie de la Digestion...Rédigées
par É. Levier. 2 tomes. 8vo. *Florence*, 1867. 125

**Schlegel** (H.). Essay on the Physiognomy of Serpents. Transl. by
Thomas S. Traill. 8vo. *Edinburgh*, 1843. 106

**Schleicher** (August). Darwinism tested by the Science of Language.
Transl....by Alex. V. W. Bikkers. 8vo. *London*, 1869. 10

**Schmid** (Rudolf). Die Darwin'schen Theorien und ihre Stellung zur
Philosophie, &c. 8vo. *Stuttgart*, 1876. 39

**Schmidt** (Oscar). Descendenzlehre und Darwinismus. 8vo. *Leipzig*,
1873. 13
The doctrine of Descent and Darwinism. 8vo. *London*, 1875. 11

**Schneider** (Georg H.). Der thierische Wille. 8vo. *Leipzig*, 1880. 23

**School of Mines.** *See* Museum of Practical Geology, &c. Ng

**Schopenhauer** (Arthur). *See* Busch (Otto). 113

**Schübeler** (F. C.). Die Pflanzenwelt Norwegens. 4to. *Christiania*,
1873–75. Nc

**Schultze** (Fritz). Kant und Darwin. 8vo. *Jena*, 1875.    39
Die Sprache des Kindes. (Darwinistische Schriften, Nr. 10.) 8vo.
*Leipzig*, 1880.    41

**Schulz** (Ernst). Nine Plates of Photographs.—Facial Expression. 4to.
   **Na**

**Schwann** (Th.). Manifestation en l'honneur de M. le Professeur Th.
Schwann. Liége, 23 juin 1878. Liber Memorialis. 8vo. *Düsseldorf*, 1879.    123

**Schwarz** (Eduard). *See* Weisbach (A.). Reise der österr. Fregatte
Novara. Anthrop. Theil, 2te Abth. 4to. *Wien*, 1867.    67

**Schweigger** (*Prof.*). *See* Grant (R. E.). Observations on...Silicious
Sponges. 8vo. *Edinburgh*.    11

**Sclater** (Philip L.). *See* Nitzsch (C. L.). Pterylography. (Ray Soc.
Publ.) Fol. *London*, 1867.    72

**Scoresby** (W.), *Jun.* An account of the Arctic Regions, &c. 2 vols.
8vo. *Edinburgh*, 1820.    15

**Scott** (John). *See* Douglas (J. W.). The British Hemiptera. Vol. 1.
8vo. *London*, 1865.    17
Report on the experimental culture of the Opium Poppy...for the
season ending 15th April, 1874. Fol. *Calcutta*, 1874.    **Nc**
do. For the season ending 31st May, 1875. Fol. *Calcutta*, 1876. **Nc**
do. For the season 1877–78. Fol. *Calcutta*, 1878.    **Nc**
Manual of Opium Husbandry. 8vo. *Calcutta*, 1877.    60

**Scott** (W. R.). The Deaf and Dumb. 2nd ed. 8vo. *London*, 1870. 14

**Scrope** (G. P.). Considerations on Volcanos. 8vo. *London*, 1825. 113
Volcanos. The Character of their Phenomena, &c. 8vo. *London*,
1862.    113

**Scudder** (John M.). Specific Diagnosis. 8vo. *Cincinnati*, 1874.    104

**Scudder** (Samuel H.). Historical sketch of the generic names proposed
for Butterflies. (Extr.) 8vo. *Salem*, 1875.    94
Butterflies. 8vo. *New York*, 1881.    102
*See* Harris (T. W.).    102

**Sedgwick** (*Rev.* Adam), *Woodwardian Prof.* A discourse on the Studies
of the University of Cambridge. 5th ed. 8vo. *London*, 1850.    14
Geology of the Lake District of Cumberland, Westmorland and Lancashire.
8vo. *Kendal*, 1853.    98
*See* Seeley (H. G.). Index to the fossil remains of Aves, &c. 8vo.
*Cambridge*, 1869.    97
Supplement to the Memorial of the Trustees of Cowgill Chapel. (Pr.
pr.) 8vo. *Cambridge*, 1870.    113

**\*Sedgwick** (Adam). *See* Cambridge.    **Ng**

**Seeley** (Harry Govier). Index to the fossil remains of Aves, Ornithosauria, and Reptilia, from the Secondary System of Strata...With a Prefatory Notice by the Rev. Adam Sedgwick. 8vo. *Cambridge,* 1869. **97**
The Ornithosauria. 8vo. *Cambridge,* 1870. **107**

**Seemann** (Berthold). Flora Vitiensis. (Fiji Islands.) 4to. *London,* 1865–73. **Nc**

**Seguenza** (G.). Ricerche paleontologiche intorno ai Cirripedi terziarii della Provincia di Messina. Parte 1, 2. 4to. *Napoli,* 1874–76. **74**

**Seidlitz** (Georg). Die Darwin'sche Theorie. Elf Vorlesungen. 8vo. *Dorpat,* 1871. **39**
do. 2te Aufl. 8vo. *Leipzig,* 1875. **39**
Beiträge zur Descendenz-Theorie. 8vo. *Leipzig,* 1876. **39**

**Selby** (Prideaux J.). Ornithology. Pigeons. (The Naturalist's Library, Vol. IX.) 8vo. *Edinburgh,* n. d. **118**

**Semper** (Karl). Die Palau-Inseln im stillen Ocean. 8vo. *Leipzig,* 1873. **15**
Die Verwandtschaftsbeziehungen der gegliederten Thiere. (Extr.) 8vo. *Hamburg,* 1876. **96**
The natural conditions of existence as they affect Animal Life. 8vo. *London,* 1881. **11**

**\*Semple** (C. E. A.). Elements of Materia Medica and Therapeutics. 8vo. *London,* 1892. **104**

**Settegast** (H.). Die Thierzucht. 8vo. *Breslau,* 1868. **Nf**

**\*Shaftesbury** (Anthony, *3rd Earl of*). Characteristicks of Men, Manners, Opinions, Times. 3 vols. (Reprint.) 12mo. 1749. **118**

**\*Sharp** (David). Fauna Hawaiiensis. Vol. 3, Part 3, Coleoptera. ii. 4to. *Cambridge,* 1903. **Nb**

**Sharpe** (William), *M.D.* Man a special creation. 8vo. *London,* 1873. **10**

**\*Shaw** (James). *See* Wallace (Robert). **113**

**Shirreff** (Patrick). Improvement of the Cereals and an Essay on the Wheat-Fly. (Pr. pr.) 8vo. *Edinburgh,* 1873. **24**

**Shuckard** (W. E.). Essay on the indigenous fossorial Hymenoptera. 8vo. *London,* 1837. **102**

**Sidgwick** (Henry). The Methods of Ethics. 8vo. *London,* 1874. **22**
A Supplement to the first edition of the Methods of Ethics. 8vo. *London,* 1877. **22**

**Sidney** (Samuel). *See* Youatt (Wm.). The Pig. 8vo. *London,* 1860. **118**

**Siebold** (C. Th. v.). The Progress of Zoology in 1842. Transl. by W. B. Macdonald. (Ray Soc. Publ.—Reports...1841, 1842). 8vo. *London*, 1845.     **17**
Reports on Zoology for 1843, 1844. Transl. (Ray Soc. Publ.) 8vo. *London*, 1847.     **17**
Nouveau Manuel d'Anatomie comparée. Par C. T. de S., et H. Stannius. Trad. par A. Spring et Th. Lacordaire. 2 tomes (in 3). 8vo. *Paris*, 1850.     **128**
On a true parthenogenesis in Moths and Bees. Transl. by Wm. S. Dallas. 8vo. *London*, 1857.     **94**
Beiträge zur Parthenogenesis der Arthropoden. 8vo. *Leipzig*, 1871.     **106**

**Siegwart** (Karl). Das Alter des Menschengeschlechts. 3te Separat-Ausgabe. 8vo. *Berlin*, 1873.     **10**

**Silas** (Ferdinand). *See* Scherzer (C. de). La Province de Smyrne. Transl. 8vo. *Vienne*, 1873.     **15**

**Sitzungsberichte der k. k. Akademie der Wissenschaften.** Mathematisch-Naturwissenschaftliche Classe. Bde 63; 65; 67; 68, iii–v; 69, iii–v; 70; 71, iii–v; 72, iv, v; 73; 74, iii–v; 76–80; 81, i–iv; 82, i, ii; 83, i–iv. 8vo. *Wien*, 1871–81.     **42**

**Skertchly** (Sydney B. J.). The Physical System of the Universe. 8vo. *London*, 1878.     **9**

**Sketch** (A) of a Philosophy. Part 2. Matter and Molecular Morphology. 8vo. *London*, 1868.     **122**

**Smellie** (William). The Philosophy of Natural History. 4to. *Edinburgh*, 1790.     **8**

**Smith** (Alexander). The Philosophy of Morals. 2 vols. 8vo. *London*, 1835.     **22**

**Smith** (Andrew). Illustrations of the Zoology of South Africa. 4 vols. 4to. *London*, 1849.     **67**

**Smith** (*Lieut.-Col.* Chas. H.). Dogs. 2 vols. (The Naturalist's Library, vols. 9, 10.) 8vo. *Edinburgh*, 1839–40.     **118**
Horses. (The Naturalist's Library, vol. 12.) 8vo. *Edinburgh*, 1841.     **118**
The Natural History of the Human Species. 8vo. *London*, 1852.     **114**

**Smith** (Frederick). *See* British Museum. Catalogue of British Hymenoptera. Part 1. 8vo. *London*, 1855.     **96**

**Smith** (*Sir* James E.). A Grammar of Botany. 8vo. *London*, 1821.     **61**
The English Flora. 4 vols. 8vo. *London*, 1824–28.     **61**

**Smith** (J. Toulmin). The Ventriculidæ of the Chalk. 8vo. *London*, 1848.     **97**

**Snell** (Karl). Die Schöpfung des Menschen. 8vo. *Leipzig*, 1863.     **9**

**Sole** (Francesco). Il Positivismo. 8vo. *Napoli,* 1881.    **28**
Su la Sensazione. 8vo. *Napoli,* 1882.    **28**

**Solis y Rivadeneyra** (Antonio de). Historia de la Conquista de Mexico. 4to. *Madrid,* 1790.    **15**

**Solms-Laubach** (*Graf* zu). Fauna und Flora des Golfes von Neapel. 4. Monographie: Corallina. 4to. *Leipzig,* 1881.    **Na**

**Somerville** (Mary). On the connexion of the Physical Sciences. 8vo. *London,* 1834.    **26**
On molecular and microscopic Science. 2 vols. 8vo. *London,* 1869.    **106**

**Soret** (J. Louis). François Jules Pictet. Notice biographique. 8vo. *Genève,* 1872. (Extr.)    **113**

**Soury** (Jules). *See* Haeckel (E.). Les preuves du Transformisme. Trad. 8vo. *Paris,* 1879.    **11**

**Sowerby** (John E.). British Poisonous Plants. Illustr. by J. E. S. Described by Ch. Johnson and C. P. Johnson. 2nd ed. 8vo. *London,* 1861.    **62**

**Spence** (William). *See* Kirby (*Rev.* Wm.).    **112**

**Spencer** (Herbert). The Principles of Psychology. 8vo. *London,* 1855.    **23**
do. 2nd ed. 2 vols. 8vo. *London,* 1870–72.    **28**
First Principles. (Subscriber's Copy.) 8vo. *London,* 1860–62.    **23**
*do. 3rd ed. 8vo. *London,* 1870.    **23**
*do. 6th ed. 8vo. *London,* 1900.    **23**
*Education. 8vo. *London,* 1861.    **22**
Essays: scientific, &c. (2nd series.) Repr. 8vo. *London,* 1863.    **39**
The Classification of the Sciences. 3rd ed. 8vo. *London,* 1871.    **39**
Descriptive Sociology. Classified and arranged by H. S....English. Compiled and abstr. by James Collier. Fol. *London,* 1873.    **Q. 2**
Grundlagen der Philosophie. Autorisirte deutsche Ausg....von B. Vetter. 8vo. *Stuttgart,* 1875.    **23**
The Principles of Sociology. Vol. 1. 8vo. *London,* 1876.    **23**
The Study of Sociology. 6th ed. 8vo. *London,* 1877.    **11**
*The Factors of Organic Evolution. (Reprint.) 8vo. *London,* 1887.    **23**

**Spiritual Evolution.** An Essay on Spiritual Evolution...By J. P. B. 8vo. *London,* 1879.    **28**

**Spix** (Joh. Bapt. von) and C. F. Phil. **von Martius.** Travels in Brazil in the years 1817–1820. 2 vols. 8vo. *London,* 1824.    **15**

**Sprengel** (Christian Konrad). Das entdeckte Geheimniss der Natur im Bau und in der Befruchtung der Blumen. 4to. *Berlin,* 1793.    **44**

**Sprengel** (J. W.). Die Fortschritte des Darwinismus. (Separatabdruck.) 8vo. *Cöln,* 1874.    **9**

**Spring** (A.) et **Lacordaire** (Th.).  *See* Siebold (C. Th. v.).  128

**\*Squire** (Balmanno).  A manual of the diseases of the Skin.  3rd smaller ed.  8vo. *London,* 1885.  104

**Stainton** (H. T.).  A manual of British Butterflies and Moths.  2 vols. 8vo. *London,* 1857–59.  102
*See* Buckler (Wm.).  The larvæ of the British Butterflies and Moths. 2 vols.  8vo. *London,* 1886–87.  17a

**Stannius** (H.).  *See* Siebold (C. Th. v.).  128

**Stebbing** (Thomas R. R.).  Essays on Darwinism.  8vo. *London,* 1871. 10

**Steenstrup** (Joh. J. Sm.).  On the alternation of Generations.  Transl. ...by G. Busk.  (Ray Soc. Publ.) 8vo. *London,* 1845.  17
Hectocotyldannelsen hos Octopodslægterne *Argonauta* og *Tremoctopus.* (Extr.)  4to. *Kjöbenhavn,* 1856.  **Ne**

**Steenstrup** (Joh. J. Sm.) og Chr. Fred. **Lütken.**  Bidrag til Kundskab om det aabne Havs Snyltekrebs og Lernæer samt om...parasitiske Copepoder. (Extr.)  4to. *Kjöbenhavn,* 1861.  **Ne**

**Stephens** (James F.).  Illustrations of British Entomology....Haustellata. 4 vols.  (Bd. in 3.)  8vo. *London,* 1828–34.  102
A systematic catalogue of British Insects.  2 Parts (in 2 vols., interleaved).  8vo. *London,* 1829.  102
\*A Manual of British Coleoptera, or Beetles.  8vo. *London,* 1839.  102

**Sterne** (Carus), *pseud.* = Ernst Krause.  Werden und Vergehen.  8vo. *Berlin,* 1876.  26
do.  2te Aufl.  8vo. *Berlin,* 1880.  26

**\*Stirling** (J. H.).  *See* Cupples (G.).  Scotch Deer-hounds....Biographical Sketch of the author by J. H. S.  8vo. *Edinburgh,* 1894.  **Nf**

**Stonehenge.**  *Pseud.,* = John Henry Walsh.  The Dog.  8vo. *London,* 1867.  108

**Strasburger** (Ed.).  Zellbildung und Zelltheilung.  2te Aufl.  8vo. *Jena,* 1876.  122
do.  3te Aufl.  8vo. *Jena,* 1880.  122
Sur la formation et la division des Cellules.  Éd. revue et corr.  Trad. de l'Allemand par J. J. Kickx.  8vo. *Jena,* 1876.  60
\*Ueber Kern- und Zelltheilung im Pflanzenreiche, &c.  (Histologische Beiträge, Heft 1.)  8vo. *Jena,* 1888.  122
\*Ueber das Wachsthum vegetabilischer Zellhäute.  (Histologische Beiträge, Heft 2.)  8vo. *Jena,* 1889.  122

**\*Strauss** (David Friedrich).  Der alte und der neue Glaube.  2te Aufl. 8vo. *Leipzig,* 1892.  12

**Stricker** (S.).  Handbuch der Lehre von den Geweben des Menschen und der Thiere.  Hrsg. von S. S.  2 Bde.  8vo. *Leipzig,* 1871–72.  115

**Strickland** (H. E.). *See* Agassiz (L.). Bibliographia Zoologiæ et Geologiæ. Vol. 2, 3. 8vo. *London*, 1852–54.    **17**

*****Struthers** (John). Memoir on the Anatomy of the Humpback Whale, *Megaptera longimana.* (Repr.) 8vo. *Edinburgh*, 1889.    **106**

**Strzelecki** (P. E. de). Physical description of New South Wales and Van Diemen's Land. 8vo. *London*, 1845.    **25**

**Sturm** (K. Ch. G.). Ueber Raçen...der landwirthschaftlichen Haus-thiere. Hrsg. von K. Ch. G. S. 8vo. *Eberfeld*, 1825.    **108**

**Sully** (James). Sensation and Intuition. 8vo. *London*, 1874. (2 copies.)    **22**

**Supernatural** (The) in Nature. 8vo. *London*, 1878.    **12**

**Supernatural Religion.** 2nd ed. 2 vols. 8vo. *London*, 1874. [By — Cassels. Ascribed by some to Dr John Muir.]    **12**

**Survival** (The). With an apology for Scepticism. 8vo. *Lond.*, 1877.    **22**

**Swainson** (William). *See* Richardson (J.). Fauna Boreali-Americana. 3 Parts. 4to. *London*, 1829–36.    **75**
A treatise on the Geography and Classification of Animals. (Lardner's Cab. Cyclop.) 8vo. *London*, 1835.    **128**
On the Natural History and Classification of Birds. 2 vols. (Lardner's Cab. Cyclop.) 8vo. *London*, 1836–37.    **128**

**Swank** (James M.). Statistics of the Iron and Steel Production of the United States. 4to. *Washington*, 1881.    **74**

**Swinhoe** (Robert). Narrative of the North China Campaign of 1860. 8vo. *London*, 1861.    **15**
Notes on the Island of Formosa. [Extracts from various Periodicals.] 8vo. *Newcastle and London*, 1863.    **16**

**Sykes** (*Major* W. H.). A Catalogue of the Mammalia observed in Dukhun, East Indies. (Extr.) 8vo. *London*, 1831. [Philos. Tracts, i. 14.]    **11**

**Syme** (Patrick). *See* Werner's Nomenclature of Colours. 2nd ed. 8vo. *Edinburgh*, 1821.    **24**

**T. E. S. T.** *See* Two Kinds of Truth. 8vo. *London*, 1890.    **28**

**Tayler** (John James). Christianity : what is it ? and what has it done ? 8vo. *London*, 1868.    **14**

*****Taylor** (Hugh). The Morality of Nations. 8vo. *London*, 1888.    **22**

**Taylor** (J. E.). Flowers ; their origin, &c. 8vo. *London*, 1878.    **59**

**Taylor** (Richard). Scientific Memoirs, selected from the Transactions of Foreign Academies of Science, &c. Ed. by R. T. Vol. 1. 8vo. *London*, 1837.    **40**

**Teale** (T. Pridgin). Dangers to Health. 8vo. *London*, 1878.    **14**

**Tegetmeier** (W. B.). Profitable Poultry. New ed. 8vo. *London,*
1854. 127
The Poultry Book : including Pigeons and Rabbits (by Harrison Weir).
Parts I–XI. (Incomplete.) 4to. *London,* 1856–57. 75
The Poultry Book. 4to. *London,* 1867. 75
Pheasants for Coverts and Aviaries. 4to. *London,* 1873. 74
*See* Blyth (Edw.). The Natural History of the Cranes. 8vo. *London,*
1881. **Nf**

**Thiselton-Dyer** (*Sir* W. T.). *See* Dyer (W. T. Thiselton).

**Thomas** (Cyrus). Synopsis of the Acrididae of North America. [U.S.
Geological Survey.] 4to. *Washington,* 1873. 74
*See* United States Entomological Commission. 2nd Report, 1878–79.
8vo. *Washington,* 1880. 102

**Thompson** (J ). A new...Ready Reckoner. 8vo. *Gainsborough,* 1805.
118

**Thompson** (Wm.). The Natural History of Ireland. Vols. 1–3. Birds.
8vo. *London,* 1849–51. 16

**Thomson** (David). Handy Book of the Flower-Garden. 2nd ed. 8vo.
*Edinburgh,* 1871. 59

**\*Thomson** (J. A.). The Science of Life. 8vo. *London,* n. d. 59
\*The Study of Animal Life. 8vo. *London,* 1892. 108

**Thomson** (Thomas). *See* Hooker (J. D.). Flora Indica. Vol. 1.
8vo. *London,* 1855. 59

**Thorell** (T.). On European Spiders. Part 1. 4to. *Upsala,* 1869–70.
**Ne**
Remarks on Synonyms of European Spiders. Nos. 1–4. 8vo. *Upsala*
(1870–73). **Ne**
Études scorpiologiques. 8vo. *Milan,* 1877. 106

**Thoughts on the Mental Functions**. 8vo. *Edinburgh,* n. d. 12

**Tieghem** (Ph. van). Traité de Botanique. Fasc. 1–4. 8vo. *Paris,*
n. d. 58
*See* Sachs (Julius). Traité de Botanique. 8vo. *Paris,* 1874. 58

**Tietze** (Emil). Ueber die Devonischen Schichten von Ebersdorf. 4to.
*Cassel,* 1870. 74

**Timiriazeff** (C.). (An Essay on the Theory of Darwin. In Russian.)
8vo. *St Petersburg,* 1865. 23

**Todd** (Robert B.). *See* Cyclopædia (The) of Anatomy and Physiology.
5 vols. 8vo. *London,* 1859. **Ng**

**\*Tollens** (B.). Kurzes Handbuch der Kohlenhydrate. 8vo. *Breslau,*
1888. 105

**Tornøe** (Hercules). Chemistry. (The Norwegian North-Atlantic Expedition, 1876–78.) 4to. *Christiania,* 1880. 72

**Tournouër** (R.). *See* Gaudry (A.). Animaux fossiles du mont Léberon. 4to. *Paris*, 1873. **Nb**

**Traill** (Thos. S.). *See* Schlegel (H.). **106**

**Trémaux** (P.). Origine et transformations de l'Homme, &c. 1ère Partie. 8vo. *Paris*, 1865. [2 copies.] **11**

**Treub** (Melchior). Notes sur l'embryogénie de quelques Orchidées. (Extr.) 4to. *Amsterdam*, 1879. **Nb**

**Trimen** (Henry). Flora of Middlesex. By H. T., and W. T. Thiselton Dyer. 8vo. *London*, 1869. **62**

**Troschel** (F. H.). Reports on Zoology for 1843–44. Transl. (Ray Soc. Publ.) 8vo. *London*, 1847. **17**
The Progress of Zoology in 1842. Transl. by W. B. Macdonald. (Ray Soc. Publ.—Reports...1841–42.) 8vo. *London*, 1845. **17**

**Tschudi** (Friedrich von). Sketches of Nature in the Alps. From the German. 8vo. *London*, 1856. **25**

**Tulk** (Alfred). *See* Wagner (R.). Elements of the comparative anatomy of the Vertebrate Animals. 8vo. *London*, 1845. **115**

**Tullberg** (Tycho). *See* Linnæus. **Na**

**Turner** (William). *See* Paget (J.). Lectures on Surgical Pathology. 3rd ed. 8vo. *London*, 1870. **93**

**Turton** (W.). British Fauna...arranged according to the Linnean System. Vol. 1. 12mo. *Swansea*, 1807. **106**

**Tuttle** (Hudson). The origin and antiquity of Physical Man scientifically considered. 8vo. *Boston*, 1866. **10**

**Twining** (Thomas). Science for the People. 8vo. *London*, 1870. **26**

**\*Two Kinds of Truth.** A test of all Theories. By T. E. S. T. 8vo. *London*, 1890. **28**

**Tylor** (Edward Burnet). Researches into the early history of Mankind. 8vo. *London*, 1865. **114**
Researches into the early history of Mankind. 2nd ed. 8vo. *London*, 1870. **124**
Primitive Culture. 2 vols. 8vo. *London*, 1871. **114**
Anthropology. 8vo. *London*, 1881. **114**

**Tyndall** (John). On the Physical Phenomena of Glaciers. Part 1. (Extr.) 4to. *London*, 1858. **74**
Essays on the use and limit of the Imagination in Science. 8vo. *London*, 1870. **12**
Address delivered before the British Association assembled at Belfast. With Additions. 8vo. *London*, 1874. **23**

**Ulloa** (Antonio de). *See* Juan (George). **95**

*Unger (F.). Botanische Briefe. 8vo. *Wien*, 1852.     62
Versuch einer Geschichte der Pflanzenwelt. 8vo. *Wien*, 1852.     60

United States Entomological Commission. Second Report...
1878–79, relating to the Rocky Mountain Locust and the Western
Cricket. 8vo. *Washington*, 1880.     102
Bulletin. No. 3. The Cotton Worm. By Chas. V. Riley. 8vo.
*Washington*, 1880. No. 7. Insects injurious to Forest and Shade
Trees. By A. S. Packard, *Jr.* 8vo. *Washington*, 1881.     102

Ure (Andrew). *See* Dictionary (*Chemistry*). 2nd ed. 8vo. *London*,
1823.     105

Usher (W.). *See* Morton (S. G.). Types of Mankind.     114

Vacek (Michael). Ueber österreichische Mastodonten. (Extr.) 4to.
*Wien*, 1877.     **Na**

Van Mons (J. B.). Arbres fruitiers. 2 tomes. 12mo. *Louvain*,
1835–36.     62

Varrentrapp (Georg). *See* Lucae (J. C. G.).     72

Vasseur (Gaston). Recherches géologiques sur les Terrains tertiaires de
la France occidentale. Stratigraphie. 1ère Partie : Bretagne. 8vo.
*Paris*, 1881.     50

Vaucher (J. P.). Histoire physiologique des Plantes d'Europe. 4
tomes. 8vo. *Paris*, 1841.     60

Veith (J. E.). Die Naturgeschichte der nutzbaren Haussäugethiere.
8vo. *Wien*, 1856.     118

Verhandlungen und Mittheilungen des siebenbürgischen Vereins
für Naturw. zu Hermannstadt. xv-xviii Jahrgang. 8vo. 1864–67.
    42

Verity (Robert). Changes produced in the Nervous System by Civili-
zation. 2nd ed. 8vo. *London*, 1839.     125

Verlot (B.). Sur la production et la fixation des variétés dans les plantes
d'ornement. (Extr.) 8vo. *Paris*, 1865.     57

Viardot (Louis). Libre examen. 8vo. *Paris*, 1871.     28
do. Nouv. éd. 8vo. *Paris*, 1872.     28
do. 5ème éd. 8vo. *Paris*, 1877.     28
do. 6ème éd. 8vo. *Paris*, 1881.     28

Villers (Brochant de). *See* Dictionnaire des Sciences naturelles. Planches.
Cristallographie. 8vo. *Paris*, 1816–30.     41

Vincent (Charles W.). *See* Year-Book.     26

Virchow (Rudolf). Cellular Pathology...Twenty Lectures. Transl.
from the 2nd ed. of the original, by F. Chance. 8vo. *London*, 1840.
    102
Ueber einige Merkmale niederer Menschenrassen am Schädel. (Extr.)
4to. *Berlin*, 1875.     **Ne**

**Vivisection.** Report of the Royal Commission...with Minutes of Evidence and Appendix. Fol. *London*, 1876. **Na**

**Vöchting** (Hermann). Beiträge zur Morphologie und Anatomie der Rhipsalideen. (Extr.) 8vo. *Leipzig*, 1873. **122**
Der Bau und die Entwicklung des Stammes der Melastomeen. 8vo. *Bonn*, 1875. **122**

**Vogeli** (Félix). *See* Agassiz (L.). De l'Espèce et de la Classification en Zoologie. (Traduction.) 8vo. *Paris*, 1869. **112**

**Vogt** (Carl). Lectures on Man. Ed. by James Hunt. 8vo. *London*, 1864. **124**
Mémoire sur les Microcéphales ou Hommes-Singes. 4to. (*Paris*, 1867). **Nc**
*See* Gegenbaur (C.). Manuel d'Anatomie comparée. 8vo. *Paris*, 1874. **55**
Lettres physiologiques. 1ère éd. française de l'auteur. 8vo. *Paris*, 1875. **125**

**Volz** (K. W.). Beiträge zur Kulturgeschichte. Der Einfluss des Menschen auf die Verbreitung der Hausthiere und der Kulturpflanzen. 8vo. *Leipzig*, 1852. **57**

**Voyage à l'Isle de France, à l'Isle de Bourbon, &c.** Par un Officier du Roi. 8vo. *Neuchatel*, 1773. **15**

*****Voysey** (*Rev.* Charles). Theism as a Science of Natural Theology and Natural Religion. 8vo. *London*, 1895. **22**

**Vries** (Hugo de). Over de Bewegingen der Ranken van Sicyos. (Extr.) 8vo. *Amsterdam*, 1880. **60**

**Vulpian** (A.). Leçons sur l'appareil vaso-moteur. Rédigées et publiées par H. C. Carville. 2 tomes. 8vo. *Paris*, 1875. **94**

**Wagner** (Andr.). Reports on the Progress of Zoology in 1842. Transl. by W. B. Macdonald. (Ray Soc. Publ.—Reports...1841–42.) 8vo. *London*, 1845. **17**
Reports on Zoology for 1843–44. Transl. (Ray Soc. Publ.) 8vo. *London*, 1847. **17**

**Wagner** (Moritz). The Darwinian Theory and the law of the Migration of Organisms. Transl. by J. J. Laird. 8vo. *London*, 1873. **39**

**Wagner** (Rudolph). Elements of the comparative anatomy of the Vertebrate Animals. Ed. from the German by A. Tulk. 8vo. *London*, 1845. **115**
Die Forschungen über Hirn- und Schädelbildung des Menschen, &c. (Extr.) 4to. *Göttingen*, 1861. **44**
*See* Blumenbach (J. F.). The Anthropological Treatises of J. F. B. (On the Anthropological Collection of the Physiological Inst. of Göttingen. By R. W.) Transl. 8vo. *London*, 1865. **114**

**Waitz** (Theodor). Introduction to Anthropology. Ed....by J. F. Collingwood. 8vo. *London*, 1863.                                    124

**Wake** (C. Staniland). Chapters on Man. 8vo. *London*, 1868. [2 copies.]
                                                                10

**Walckenaer** (C. A.). *See* Azara (F. de). Voyages dans l'Amérique
méridionale...1781–1801. Tomes 1–4 and Atlas. 8vo. *Paris*, 1809.
                                                           15 & 72

**Waldner** (Heinrich). Deutschlands Farne. Heft 1, 2. Fol. *Heidelberg*, 1879–80.                                                  **Na**

**Walker** (Alexander). Intermarriage. 8vo. *London*, 1838.        10

**Walker** (Francis). Monographia Chalciditum. 8vo. *London*, 1839.  96

**\*Walker** (J. and C.). *See* Atlas (England and Wales). Fol. *London*,
1837.                                                          **Q. 1**

**Wall** (Abiathar). *See* Macaulay (J.). Vivisection. 8vo. *Lond.*, 1881.  28

**Wallace** (Alfred Russel). The Malay Archipelago. 2 vols. 8vo.
*London*, 1869.                                                  89
Contributions to the Theory of Natural Selection. 8vo. *Lond.*, 1870.  23
do. 2nd ed. 8vo. *London*, 1871.                                23
The geographical distribution of Animals. 2 vols. 8vo. *Lond.*, 1876.  41
Tropical Nature, and other Essays. 8vo. *London*, 1878.          89
Island Life. 8vo. *London*, 1880.                               89
\*Darwinism. 8vo. *London*, 1889.                               40

**\*Wallace** (Robert). A Country Schoolmaster, James Shaw. 8vo.
*Edinburgh*, 1899.                                              113

**Wallich** (G. C.). Eminent Men of the Day. Photographed by G. C. W.
Scientific Series. 4to. *London*, 1870.                        113

**Walsh** (J. H.). *See* Stonehenge.                            108

**Walther** (Alexander). Die Laubmoose Oberfrankens. Von A. W., und
L. Molendo. 8vo. *Leipzig*, 1868.                              118

**Walther** (Friedrich L.). Der Hund. 8vo. *Giessen*, 1817.     118
Das Rindvieh. 8vo. *Giessen*, 1817.                            108

**Wanderings through the Conservatories at Kew.** 8vo. *London*,
n. d.                                                           62

**Ward** (Robert A.). A treatise on Investments. 8vo. *London*, 1852.  24

**Warner** (Francis). Physical Expression; its modes and principles.
8vo. *London*, 1885.                                            11

**Waterhouse** (George R.). Mammalia (Zoology of...H.M.S. Beagle,
Part 2). 4to. *London*, 1839. [2 copies.]                       67
Marsupialia or pouched animals. (The Naturalist's Library, Vol. 11.)
8vo. *Edinburgh*, 1841.                                         118
A Natural History of the Mammalia. 2 vols. 8vo. *London*, 1846–48.
                                                                118

**Waterton** (Charles). Essays on Natural History. 2nd ed. 8vo. *London*, 1838. **118**

**Watson** (Hewett C.). The geographical distribution of British Plants. 3rd ed. Part I. 8vo. *London*, 1843. **59**
Cybele Britannica. Vols. 1–4. 8vo. *London*, 1847–59. **59**
do. Part First of a Supplement. (Pr. pr.) 8vo. *London*, 1860. **59**
do. Compendium. Parts 1–3. (Pr. pr.) 8vo. *Thames Ditton*, 1868–70. **59**

***Watson** (Thomas). Lectures on the principles and practice of Physic. 4th ed. 2 vols. 8vo. *London*, 1857. **93**

**Watts** (Henry). *See* Dictionary (*Chemistry*). 5 vols. and Supplement. 2nd ed. 8vo. *London*, 1871–72. **105**

**Webb** (Henry). Dogs; their points, &c. Ed. by H. W. 8vo. *London*, (1876). **118**

**Weber** (D. A.). Der Taubenfreund. 2te Aufl. 8vo. *Quedlinburg*, 1850. **127**

***Webster** (A. D.). British Orchids. 8vo. *Bangor*, 1886. **62**

**Weddell** (H. A.). Voyage dans le Nord de la Bolivie, &c. 8vo. *Paris*, 1853. **9**

**Wedgwood** (Hensleigh). On the developement of the Understanding. 8vo. *London*, 1848. **28**
On the origin of Language. 8vo. *London*, 1866. **14**
A Dictionary of English Etymology. 2nd ed. 8vo. *London*, 1872. **12**

**Weinland** (D. F.). Ueber die in Meteoriten entdeckten Thierreste. 4to. *Esslingen a/N.*, 1882. **72**

**Weisbach** (A.). Reise der österreichischen Fregatte Novara...1857–59. Anthropologischer Theil, 2te Abth., Körpermessungen...Bearb. von Dr A. W. 4to. *Wien*, 1867. **67**

**Weismann** (August). Ueber die Berechtigung der Darwin'schen Theorie. 8vo. *Leipzig*, 1868. **39**
Ueber den Einfluss der Isolirung auf die Artbildung. 8vo. *Leipzig*, 1872. **39**
Studien zur Descendenz-Theorie. 1, 2. 8vo. *Leipzig*, 1875–76. **39**
Beiträge zur Naturgeschichte der Daphnoiden. (Separatabdr.) 8vo. *Leipzig*, 1876–79. **106**
Beiträge zur Naturgeschichte der Daphnoiden. Abhandl. 2, 3, 4. 8vo. *Leipzig*, 1877. **106**
Studies in the Theory of Descent. Transl. and ed. by R. Meldola. With a Prefatory Notice by C. Darwin. 3 Parts. 8vo. *London*, 1880–82. **39**
*Essays upon Heredity, &c. Author. Transl. Ed. by E. B. Poulton, S. Schönland and Arthur E. Shipley. 8vo. *Oxford*, 1889. **13**
*The Germ-Plasm: a theory of Heredity. Transl. by W. Newton Parker and Harriet Rönnfeldt. 8vo. *London*, 1893.

**\*Wells** (William Charles). Two Essays: one upòn single vision with two eyes; the other on Dew: &c. 8vo. *London*, 1818. 24

**Werner's Nomenclature of Colours, &c.** By Patrick Syme. 2nd ed. 8vo. *Edinburgh*, 1821. 24

**West Riding Lunatic Asylum.** Medical Reports. Ed. by J. Crichton Browne. Vols. 1, 2 and 5. 8vo. *London*, 1871, 1872, 1875. 104

**Westminster Review.** N.S., Nos. 2, 3, April, July, 1852. 8vo. *London*. [Containing articles on "A Theory of Population" (No. 2) and on "The Future of Geology" (No. 3).] 119

**Westwood** (J. O.). Address on the recent progress...of Entomology. 8vo. *London*, 1835. [Philos. Tracts, i. 12.] 11
An introduction to the modern classification of Insects. 2 vols. 8vo. *London*, 1839–40. 102

**Weyenbergh** (H.). Periódico zoológico. Organo de la sociedad zoológica Argentina. Tomo 3. Entrega 4. 8vo. *Cordoba*, 1881. 96

**\*Wheatley** (Henry B.). What is an Index? 2nd ed. 8vo. *London*, 1879. 126

**Whewell** (*Rev.* Wm.). Address delivered...June 25, 1833...3rd General Meeting of the Br. Assoc. for the Advancement of Science. 8vo. *Cambridge*, 1833. [Philos. Tracts, ii. 18.] 11
History of the Inductive Sciences. 3 vols. 8vo. *London*, 1837. 40
*See* Mackintosh (*Sir* James). Dissertation on the Progress of Ethical Philosophy. Preface by the Rev. W. W. 2nd ed. 8vo. *Edinburgh*, 1837. 12

**White** (*Rev.* Gilbert). The Natural History of Selborne. New ed. 2 vols. 8vo. *London*, 1825. 10
do. New ed. with Notes by the Rev. Leonard Jenyns. 8vo. *London*, 1843. 10

**Whitney** (J. D.). The auriferous gravels of the Sierra Nevada of California. (Extr.) 4to. *Cambridge, Mass.*, 1879. 74
The Climatic Changes of later Geological Times. (Extr.) pp. 1–264. 4to. *Cambridge, Mass.*, 1880. 74

**Whitney** (Wm. Dwight). Oriental and Linguistic Studies. 8vo. *New York*, 1873. 14
The life and growth of Language. 8vo. *London*, 1875. 11

**Wichura** (Max). Die Bastardbefruchtung im Pflanzenreich. 4to. *Breslau*, 1865. 44

**Wiedersheim** (R.). *See* Ecker (A.). Die Anatomie des Frosches. 8vo. *Braunschweig*, 1864–82. 115

**Wiegmann** (A. F.). Ueber die Bastarderzeugung im Pflanzenreiche. 4to. *Braunschweig*, 1828. 57

**Wiesner** (Julius). Die heliotropischen Erscheinungen im Pflanzenreiche. 1, 2 Th. (Extr.) 4to. *Wien,* 1878–80.                         **Nc**
Das Bewegungsvermögen der Pflanzen. 8vo. *Wien,* 1881.                  **57**
*Wiesner und seine Schule…Festschrift. Von K. Linsbauer, L. Lins-
bauer, L. R. v. Portheim. Mit einem Vorworte von Hans Molisch.
8vo. *Wien,* 1903.                                                      **35**

**Wigand** (Albert). Der Darwinismus und die Naturforschung Newtons
und Cuviers. 2 Bde. 8vo. *Braunschweig,* 1874–77. [2 copies.] **39**

**Wilckens** (Martin). Die Rinderrassen Mittel-Europas. 8vo. *Wien,*
1876.                                                                  **108**
Form und Leben der landwirthschaftlichen Hausthiere. 8vo. *Wien,*
1878.                                                                  **108**
Grundzüge der Naturgeschichte der Hausthiere. 8vo. *Dresden,* 1880.
                                                                       **108**

**Wild Birds Protection.** Report from the Select Committee…Minutes
of Evidence, &c. Fol. *London,* 1873.                                  **Na**

**Williamson** (William C.). On some of the microscopical objects found
in the mud of the Levant, &c. (Extr.) 8vo. *Manchester,* 1847. **116**
On the recent Foraminifera of Great Britain. (Ray Soc. Publ.) Fol.
*London,* 1858.                                                         **72**

**Wilson** (John). British Farming. 8vo. *Edinburgh,* 1862.             **49**

**Wilson** (Owen S.). The Larvæ of the British Lepidoptera and their
food plants. 8vo. *London,* 1877.                                       **Nf**

**Winkler** (T. C.). Description de quelques nouvelles espèces de Poissons
fossiles des calcaires d'eau douce d'Oeningen. 4to. *Harlem,* 1861.
                                                                        **Ne**
Des Tortues fossiles conservées dans le Musée Teyler et dans quelques
autres Musées. 8vo. *Harlem,* 1869.                                     **75**

**Withering** (W.). A systematic arrangement of British Plants. Corr.
and condensed…by W. Macgillivray. 3rd ed. 8vo. *Lond.,* 1835. **59**

**Wollaston** (T. Vernon). Insecta Maderensia. 4to. *London,* 1854.   **74**
On the Variation of Species, &c. 8vo. *London,* 1856.                   **10**
*See* British Museum. Catalogue of the Coleopterous Insects of Madeira.
8vo. *London,* 1857.                                                   **102**

**Wolstein** (Johann G.). Ueber das Paaren und Verpaaren der Menschen
und der Thiere. 3te Aufl. 8vo. *Altona,* 1836.                           **9**

**Woodward** (Henry). *See* Salter (J. W.).                             **97**

**Woodward** (S. P.). A manual of the Mollusca. Parts 1–3. 8vo.
*London,* 1851–56.                                                     **117**

**Wright** (Chauncey). Darwinism (Repr.). 8vo. *London,* 1871.         **39**
Philosophical Discussions. With a biographical sketch of the Author
by Charles E. Norton. 8vo. *New York,* 1877.                           **40**

*Wünsche (Otto). Schulflora von Deutschland. 2te Aufl. 8vo. *Leipzig*, 1877. **61**

Würtenberger (Leopold). Studien über die Stammesgeschichte der Ammoniten. (Darwinistische Schriften, Nr. 5.) 8vo. *Leipzig*, 1880. [2 copies.] **41**

Wundt (Wilhelm). Nouveaux éléments de Physiologie humaine. Trad. par le Dr Bouchard. 8vo. *Paris*, 1872. **115**
*Grundzüge der physiologischen Psychologie. 2 Bde. 2te Aufl. 8vo. *Leipzig*, 1880. **115**

Yarrell (William). A History of British Fishes. 2 vols. 8vo. *London*, 1836. **106**
A History of British Birds. Vol. 1. 8vo. *London*, 1839. **127**

Year-Book (The) of facts in Science and the Arts for 1875. Ed. by Charles W. Vincent. 8vo. *London*, 1876. **26**

Youatt (W.). Cattle: their breeds, &c. (Library of Useful Knowledge.) 8vo. *London*, 1834. **108**
Sheep: their breeds, &c. (Library of Useful Knowledge.) 8vo. *London*, 1837. **108**
The Dog. 8vo. *London*, 1845. **118**
The Pig. Enlarged and re-written by Samuel Sidney. 8vo. *London*, 1860. **118**

Young (Thomas). A course of Lectures on Natural Philosophy and the Mechanical Arts. 2 vols. 4to. *London*, 1807. **8**

Zerffi (G. G.). A manual of the historical development of Art. 8vo. *London*, 1876. **14**

Ziegler (Martin). Atonicité et Zoïcité. 8vo. *Paris* (1874). **106**

Zoological Record. Vols. 1–19. 8vo. *London*, 1864–82. **51**

Zoological Society of London. Proceedings, 1830–81. **Ni, Nk**
do. Index, 1830–47, 1848–60, 1861–70. 8vo. **Ni, Nk**
Transactions. 4to. Vol. 5, Parts 2–end; 6, 7; 8, Parts 1–7, 9; 9, 10; 11, Parts 1–6. 1863–. General Index to Vols. 1–10. **64**

Zuccarini (J. G.). On the Morphology of the Coniferæ. Transl. by George Busk. (Ray Soc. Publ.—Reports and Papers on Botany.) 8vo. *London*, 1846. **17**

Zuckerkandl (E.). Reise der österreichischen Fregatte Novara... 1857–59. Anthropologischer Theil, 1te Abth., Cranien der Novara-Sammlung. 4to. *Wien*, 1875. **67**

CAMBRIDGE: PRINTED BY JOHN CLAY, M.A. AT THE UNIVERSITY PRESS.

Printed in the United States
By Bookmasters